Chord Transformations in Higher-Dimensional Networks

Chord Transformations in Higher-Dimensional Networks proposes an in-depth formal framework for generalized Tonnetze. It takes an algebraic approach and studies systems of k-chords in n-TET scales derived from a given k-mode (array of step intervals) through mode permutations and chord root translations, by combining key ideas of the neo-Riemannian Tonnetz theories with serial approaches to chordal structures. In particular, it provides the generalization of the neo-Riemannian P, R, L transformations via the notion of 'drift' operator, which is the main novelty of the approach. At the same time, the book is thorough in building the formal framework covering many moments and details, with special attention to trichords and tetrachords, which allow the geometric visualization of their structure helping to understand the more abstract transformations in higher-dimensional networks.

Features

- Chord transformations are explained from a new approach, by considering the chord as a two-component entity (root and mode), which is simpler than that of the neo-Riemannian theory
- The chords transformations presented can be easily converted to computational algorithms to deal with higher-dimensional Tonnetze
- Presents the study of chords with a scope that goes from scratch up to higher levels, about to develop research works.

Chord Transformations in Higher-Dimensional Networks

Rafael Cubarsi

CRC Press
Taylor & Francis Group
Boca Raton London New York

CRC Press is an imprint of the
Taylor & Francis Group, an **informa** business
A CHAPMAN & HALL BOOK

Designed cover image: Rafael Cubarsi

First edition published 2025
by CRC Press
2385 NW Executive Center Drive, Suite 320, Boca Raton FL 33431

and by CRC Press
4 Park Square, Milton Park, Abingdon, Oxon, OX14 4RN

CRC Press is an imprint of Taylor & Francis Group, LLC

ISBN: 978-1-032-98290-8 (hbk)
ISBN: 978-1-032-98298-4 (pbk)
ISBN: 978-1-003-59796-4 (ebk)

DOI: 10.1201/9781003597964

Typeset in LM Roman
by KnowledgeWorks Global Ltd.

Publisher's Note: This book has been prepared from camera-ready copy provided by the author.

Contents

Preface

> *... ce n'est cependant que par le secours des mathématiques que mes idées se sont débrouillées, et que la lumière y a succédé à une certaine obscurité dont je ne m'apercevais pas auparavant.*
>
> Jean-Philippe Rameau[1]

Music and mathematics, although seeming like disparate fields at first glance, share a profound and intricate relationship. From the harmony of musical chords to the structure of compositions, music is deeply rooted in mathematical principles. Pioneers such as Pythagoras, who is usually credited with the discovery of the mathematical ratios underlying musical intervals, and Johann Sebastian Bach, whose compositions reflect intricate mathematical structures, exemplify the deep connection between these fields.

Understanding musical chords offers a unique and enriching perspective on the application of mathematical concepts. This book explores the mathematical foundation of musical chords, introduces the tone and chord networks, and highlights the significance of these studies as well for mathematicians as for musicians who have a good knowledge of mathematics.

A musical chord is a group of notes played simultaneously. The intervals between notes in a scale are critical in defining the structure of chords. In Western music, these intervals are often described by using

[1] *... only with the aid of mathematics did my ideas become clear and did light replace a certain obscurity of which I was unaware before.* Traité de l'harmonie réduite à ses principes naturels, 1722 (Préface).

simple ratios. For instance, the interval between the first note (the root) and the fifth note (the perfect fifth) in a diatonic scale is typically a 3:2 ratio, which is associated with the third overtone. This ratio signifies that the frequency of the fifth note is $^{3}/_{2}$ times the frequency of the root note, which played together produce a consonant sound. Similarly, the interval corresponding to a major third (the third scale degree of a diatonic scale) is a 5:4 ratio, associated with the fifth overtone. However, just tuning has been approximated by a system of equal division of the octave, and from the middle of the 18th century the 12-tone equal temperament scale was gradually adopted, so that two consecutive scale tones maintain a constant ratio. By this way music can be translated to any arbitrary tone of the scale, by changing to a new fundamental tone.

In a 12-tone scale, the simplest and most fundamental chords are the major and minor triads, each consisting of three notes. A major triad is constructed from the root, major third, and perfect fifth. The minor triad, on the other hand, consists of the root, minor third, and perfect fifth. These ratios reveal the harmonic relationships that form the basis of musical chords. However, although in tonal music chords are used to produce a harmonious sound, they can also be used with an arbitrary number of notes and intervals to trigger specific sensations.

A geometric representation of the relationships between notes in a musical scale, especially when they form chords, is the tone network or Tonnetz. It was first introduced by the 18th-century music theorist Leonhard Euler and has since become a valuable tool in both music theory and mathematics.

The Tonnetz maps notes in a two-dimensional lattice where each point represents a note, and the connections between points indicate harmonic relationships such as perfect fifths, major thirds, and minor thirds. It can be viewed as a tiling of the plane with equilateral triangles, where each vertex represents a note. This geometric structure allows us to visualize and analyze chord progressions and key changes in a clear and intuitive manner.

Beyond individual chords and their geometric representations, the study of chord progressions and relationships within a piece of music can be approached through chord networks. These networks provide a framework for understanding the transitions between chords in a composition.

A chord network is a graph where each node represents a chord and each edge represents a transition from one chord to another. Analyzing these networks involves studying the paths and cycles within the graph, which correspond to musical sequences and repeated patterns.

For musicians, chord networks provide a framework for analyzing the harmonic structure of compositions, revealing patterns and relationships that might not be immediately apparent. They also assist composers in crafting chord progressions that are harmonically coherent and offer a guide for improvising musicians, suggesting chords that can be used within a given harmonic context. From a mathematical standpoint, tone and chord networks present a rich field of study that intersects with several areas of mathematics, including graph theory, group theory, and algebraic topology.

The study of tone and chord networks is not merely an abstract exercise; it has practical implications for various fields, including computer science and cognitive science. For instance, in computational musicology, algorithms based on tone and chord networks can be developed to analyze large datasets of musical compositions. This can lead to discoveries about stylistic trends, compositional techniques, and the evolution of musical forms over time. Similarly, understanding how the brain processes and perceives music is a key area of research in cognitive science. Tone and chord networks can model the cognitive processes involved in recognizing and predicting musical patterns, contributing to our knowledge of music perception and cognition.

The exploration of tone and chord networks offers a fascinating convergence of mathematics and music. Studying these structures not only enriches the understanding of musical theory but also opens up new avenues for research and application. By delving into the mathematical principles underlying music, one can uncover deeper patterns and connections that enhance both the appreciation and creation of music. The interdisciplinary nature of this study highlights the inherent beauty and complexity of both fields, showcasing the power of mathematics to illuminate the intricate world of music.

Barcelona, Summer 2024.

Author Bio

Rafael Cubarsi, mathematician and physicist by training, received his PhD from the Astronomy Department of the Universitat de Barcelona in 1988 with a dissertation on Chandrasekhar Stellar Systems. He developed and conducted research at the Universitat Autònoma de Barcelona and Universitat Politèctica de Catalunya, where he used to teach, and has more than 70 published papers in peer-reviewed journals to his credit. His research focused on the fields of Astronomy & Astrophysics and Mathematical Biology. Recently his interest is centered on Mathematical Theory of Music.

CHAPTER 1

Introduction

1.1 TONES AND NOTES

Notation, definitions, and basic concepts will be introduced first. A ratio of 2 between two *tones* (meaning frequencies) corresponds to the range of one octave, the ratio the ear associates with maximum consonance. For any tone $\nu \in \Omega \equiv (0, \infty)$, the values $2^k\nu$, $k \in \mathbb{Z}$, define one equivalence class. The set Ω is a commutative group for multiplication. The set of all the octaves of the fundamental frequency ratio (it can be assumed $\nu_0 = 1$) is a monogenous subgroup of Ω of infinite cardinal, $\Omega^2 = \{2^k, k \in \mathbb{Z}\}$. The *frequency classes* (FC) are the elements of the quotient group $\Omega_0 = \Omega/\Omega^2$, also commutative for multiplication. For each equivalence class, we may choose one representative tone in $[1, 2)$ (with identified extremes). This is the reference octave that can be identified with Ω_0. A finite set of these representative tones is referred to as a *scale*.

The following notation is useful. Given a tone $\nu \in \Omega$, its class $\overline{\nu} \in \Omega_0$ can be notated by using the concept of projection function [13] that maps[1] any frequency ratio $\nu \in \Omega$ to the FC $\pi_\nu = \overline{\nu} \in \Omega_0$.

Alternatively, each FC π_ν in the frequency space is associated with a *note* or pitch class (PC) $\log_2 \pi_\nu$ in the corresponding octave $S_0 = \mathbb{R}/\mathbb{Z}$. Ratios between tones in Ω_0 are associated with intervals in S_0. Sometimes it is more useful to work in the multiplicative space of tones and sometimes in the additive space of notes.

[1]In a more formal context, we consider a GIS [25, 26] $\pi : \Omega \times \Omega \to \Omega_0$ and a canonical GIS $\pi : \Omega_0 \times \Omega_0 \to \Omega_0$, with the α-projection function defined as $\pi_\alpha(\cdot) \equiv \pi(\alpha, \cdot)$.

DOI: 10.1201/9781003597964-1

1.2 SCALE EXAMPLES

A n-tone scale will be expressed as the set

$$E_n = \{x_k \in \Omega_0;\; k = 0, ..., n-1\}$$

for n FCs arbitrarily defined in the reference octave Ω_0.

For example, starting from the fundamental tone 1, we may select the FCs of the supersubharmonics p/q with $p, q = 1, 3, 5$, that is $\{1, \frac{6}{5}, \frac{5}{4}, \frac{4}{3}, \frac{3}{2}, \frac{5}{3}, \frac{8}{5}\}$. This is a 5-*odd-limit* tuning system, according to Harry Partch's notation.

Alternatively, starting from the fundamental tone 1 (assuming it is the first scale degree or tonic), we may select the following FCs: the third harmonic $\frac{3}{2}$ (fifth scale degree or dominant), the third subharmonic $\frac{4}{3}$ (fourth scale degree or subdominant), and for each of these classes, we may add their third and fifth harmonics (third scale degree or median). Thus, we obtain the triplets formed by the FCs $\{1, \frac{3}{2}, \frac{5}{4}\}$, $\{\frac{3}{2}, \frac{9}{8}, \frac{15}{8}\}$, and $\{\frac{4}{3}, 1, \frac{5}{3}\}$, which form the major triads[2]. So, we obtain the seven FCs $\{1, \frac{9}{8}, \frac{5}{4}, \frac{4}{3}, \frac{3}{2}, \frac{5}{3}, \frac{15}{8}\}$ forming the major *diatonic scale*, which is the basis of most of traditional melodies. This is a 5-*prime-limit* tuning system, also according to Harry Partch's notation. The inversion of the above tonal ratios leads to the minor diatonic scale. Scales, as in the previous examples, where each tone corresponds to a simple ratio between coprime harmonics, are *just intonation* scales.

Similarly, if the set $\{\frac{4}{3}, 1, \frac{3}{2}\}$ is expanded with the classes of their fifth harmonics and subharmonics, we get the 12-tone syntonic just intonation scale of Figure 1.1.

Scales can also be constructed from one generator tone. Cyclic scales generated by a real positive value h [13, 14] (also known as generalized Pythagorean scales or well-formed scales of one generator) are those obtained from the FCs of powers $\nu_0 h^k, k \in \mathbb{N}$, for certain tone ν_0 (which can be assumed as unit). By using the notation of projection functions, these FCs can be notated as $\pi_h^k = \pi_{h^k}$. In addition, they must satisfy a closure condition [3–5]. For $h = 3$ we get the particular case of the 12-tone Pythagorean scale, generated by fifths. If h is a rational power of 2, we get equal temperament scales, which are considered as degenerate cases.

[2]The term triad is used similarly to trichord, meaning a set of three notes, not necessarily in a particular segment of a scale. This meaning is extended to tetrachords, hexachords, etc.

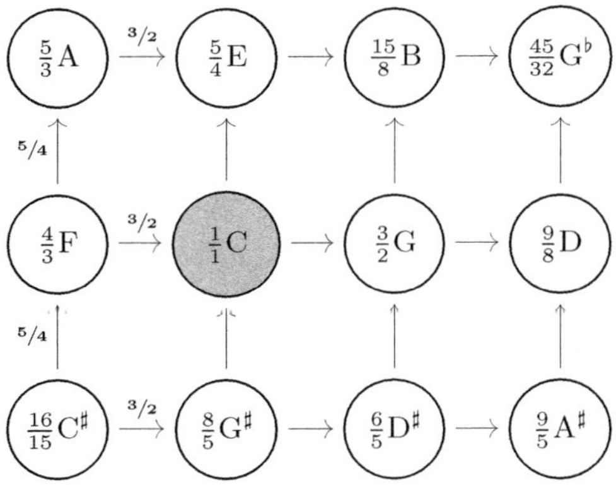

Figure 1.1 12-tone syntonic just intonation scale. Along rows, the intervals between notes are fifths, and along columns, the intervals are major thirds. Transformations operate by multiplication between pitch classes.

Thus, in an n-tone cyclic scale E_n^h the scale tones are $\nu_k \equiv \pi_h^k = \frac{h^k}{2^{[\![k]\!]}}$; $k = 0, \ldots, n-1$; with $[\![k]\!] = \lfloor k \log_2 h \rfloor$ (integer part). The FC ν_n, which does not belong to the scale E_n^h, provides the closure condition, meaning that there are no other scale tones between ν_n and the fundamental. Hence, either $\nu_n \to 1^+$ or $\nu_n \to 2^-$. When the scale tones are ordered from lowest to highest pitch in Ω_0, i.e., in *cyclic order*, we find two extreme tones, the minimum tone ν_m and the maximum tone ν_M, which indices satisfy $n = m + M$, all of them coprime.

1.3 EQUAL TEMPERAMENT SCALES

An n-tone equal temperament (n-TET) scale can be described as the following cyclic scale. The set of frequencies

$$\Omega^{2^{1/n}} = \{2^{\frac{k}{n}}, k \in \mathbb{Z}\} \tag{1.1}$$

is a subgroup of $(\Omega, \cdot)$ isomorphic to $(\mathbb{Z}, +)$. The respective FCs are the elements of the quotient group $\Omega^{2^{1/n}}/\Omega^2$, which has a structure similar to $(\mathbb{Z}_n, +)$, with classes notated as $\{2^{\frac{k}{n}}, k \in \mathbb{Z}_n\}$. Then, the n-TET scale is formed from the representatives of these FCs in Ω_0 as

$$E_n^{\top} = \{2^{\frac{k}{n}}; k = 0, \ldots, n-1\} \tag{1.2}$$

or, equivalently in S_0 (logarithmic scale of frequencies), as the PCs $\{\frac{k}{n}; k = 0, \ldots, n-1\}$.

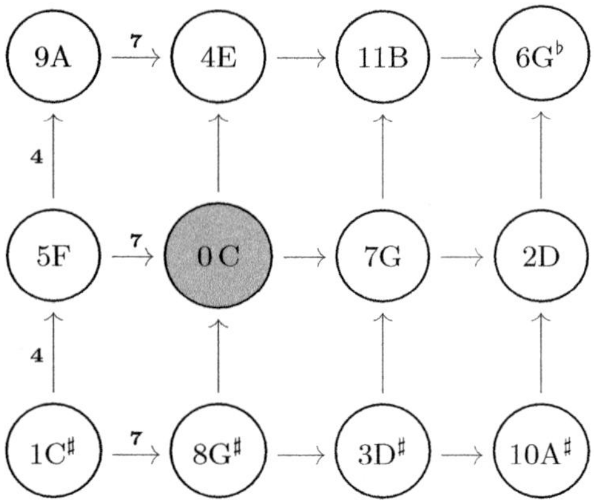

Figure 1.2 12-tone equal temperament scale. Along rows, the intervals between notes are fifths, and along columns, the intervals are major thirds. Transformations operate by addition in $\mathbb{Z}_{12}$.

Figure 1.2 shows the structure of a 12-TET scale, which is equivalent to those of syntonic just intonation and 12-tone Pythagorean scales. However, in this case, the intervals between fifths correspond to 7 semitones, and the intervals between thirds are of 4 semitones (operating in the additive space of $\mathbb{Z}_{12}$).

With reference to the 12-TET scale, an interval $\frac{1}{1200}$ in the octave S_0 is called a *cent*, obtained by dividing a semitone into 100 parts. The length of an interval with frequency ratio r is given in cents by

$$¢(r) = 1200 \log_2 r \tag{1.3}$$

which is also the melodic distance between the ends of that interval. Smaller divisions than one semitone are referred to as *microtonality*.

1.4 ADVANTAGES OF EQUAL-TEMPERAMENT

Instruments tuned so that any two consecutive scale tones maintain approximately a constant ratio have existed from long time ago. By this way, music can be translated to any arbitrary tone of the scale by changing to a new fundamental tone. This tuning system of *equal division of the octave*, ignores the exact harmonic ratios that maintain the scale tones.

In China, for instance, equally tempered pentatonic and heptatonic scales were traditionally used. Helmholtz [23] refers to Zhu Zaiyu, prince of the Ming dynasty, as the first one who calculated with accuracy the scale of 12-TET in 1580. Nearly simultaneously, in 1584, the Flemish mathematician Simon Stevin also did the same calculation, inspired by works of the lutenist and theorist Vincenzo Galilei [8].

In Western countries the discussion arose especially in the Baroque, where different musicians wrote pieces for a 12-TET scale. Musicians such as Girolamo Frescobaldi were supporters of equal temperament scales, and others, including Giuseppe Tartini, were detractors who advocated for just intonation systems. However, in the West, equal temperament tunings began to become widespread during the eighteenth century, although construction of equally tempered pianos and keyboards, as those of nowadays, did not take hold until the early twentieth century and has become the current standard.

Indeed, the small distance between their notes and those associated with simple ratios or those of the Pythagorean tuning, in case they were perceived [33], is not significant (at most, one syntonic comma or one Pythagorean comma, i.e., 21–23 ¢) in comparison to the advantages they provide. This especially concerns the notes of the diatonic scale, as shown in Figure 1.3.

Thus, precisely because the triads that form the major and minor chords do not maintain the exact ratios of just intonation, in the 12-TET system it is possible to translate them up or down without changing the musical intervals.

At the beginning of the 20th century, free atonal composition and the twelve-tone technique, either the dodecaphonic serialism or the non-dodecaphonic serial composition, ended up giving total freedom in the

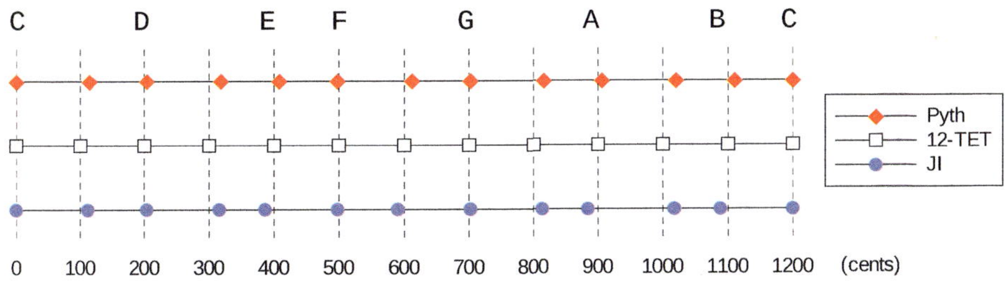

Figure 1.3 Twelve-tone scales: (top) Pythagorean starting at F, (middle) equal temperament, and (bottom) syntonic just intonation.

way of gathering notes to form chords, so that with the exception of spectral music or performances with acoustic fretless string instruments and some wind instruments with a historically informed approach based on just intonation, maintaining simple ratios between their notes is no longer a criterion to take into account, especially when it comes to music ensembles.

1.5 TONE NETWORK

In atonal music, the chords are typically characterized in terms of the intervals between their PCs. Pitch class set theory [20, 31, 32] is devoted to this. Proximity of chords is based on their distribution on the Tonnetz according to an equal temperament system, where PCs are considered instead of pitches. However, from a tonal point of view, we must remark that chord inversions have different timbres and harmonic dissonances [12], although their relative position in the network is the same.

Let us consider the example of the major triads in the 12-TET scale. Proximity between pitches, based on the harmonic distance, has the hierarchy induced by the order of the prime harmonics 3, 5, 7, 11, etc. The PC of the third harmonic – the fifth – corresponds to a note 7 semitones away from the fundamental, and the class of the fifth harmonic – the major third – is 4 semitones far from the fundamental. The ratio between these notes, $6/5$, provides the minor third, at 3 semitones from the fundamental. Thus, the multiplicative factors between FCs are associated with translations in the logarithmic equal temperament scale of PCs. The tone network of Figure 1.4 shows the triads around the C major chord $(\mathrm{C}, \mathrm{E}, \mathrm{G})$, generated by the two lowest prime harmonics other than the fundamental.

In this figure, the triangles shaped as ∇ are major chords, and those shaped as Δ are minor chords. If we focus on the *cell of chords* containing the note C, we see that adjacent chords share two notes, one of them C. This is the domain of the note C, which contains, in regard to the composition of their intervals, two different *modes* or interval arrays: three major chords with the mode $[4,3,5] = [4,3,-7]$ (measured in $\mathbb{Z}_{12}$) according to a clockwise-oriented cyclic path and three minor chords with the mode $[3,4,5] = [3,4,-7]$ according to a counterclockwise-oriented cyclic path.

The major chords are related by a simultaneous translation of their notes of 7, 4, and 3 semitones along the respective arrows from another

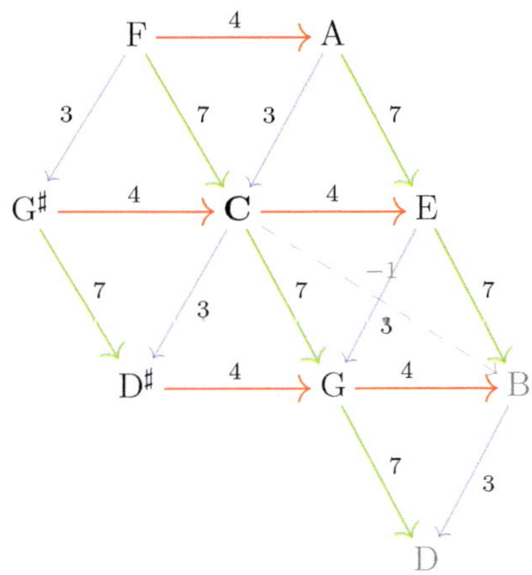

Figure 1.4 Tone network in the 12-TET scale generated by the third and fifth harmonics, centered in the note C. Color arrows indicate the number of semitones corresponding to the translations of a fifth (green), a major third (red), and a minor third (blue). The dashed arrow points to the leading tone B in a neighbor cell.

major chord, or of their opposite intervals of 5, 8, and 9 semitones in the opposite sense. The same applies to minor chords. By translations, both modes cannot be mixed. If the F major chord (F, A, C) is translated one fifth (7 semitones along the green arrow), we reach the C major chord (C, E, G). This is the direction of the harmonically closest major chord [12]. The only chord sharing two notes with the C major chord that does not belong to the chord cell of C is the E minor chord (E, G, B). This is the chord of the tone network that is at minimum harmonic distance from the C major chord. This minor chord introduces the new note B, the seventh degree of the diatonic major scale. It is called *leading tone* and is one semitone below the fundamental. For the minor diatonic scale, the seventh degree is called the subtonic. To move from the C chord cell to a cell of a note not contained in the former cell (with chords that at least share one note with the C chord cell), the less harmonically dissonant way is going through the E minor chord, which belongs to the cell of the leading tone B and is adjacent to the C major chord. If the path is limited to the use of major chords, then the chord harmonically closer to C is the G major chord (G, B, D), which also contains the leading tone and belongs to the D chord cell.

1.6 CHORD NETWORK

The analytical framework to explore the transformations between chords is the neo-Riemannian music theory [10,11,18,26]. Firstly, it was applied to tonal music, focusing on triadic harmonies, and later it was extended to atonal music as well. Originating from the work of 19th-century music theorist Hugo Riemann, this approach emphasizes voice leading, which is the smooth movement of individual notes from one chord to the next, rather than traditional harmonic functions.

To this purpose, it is more useful to use the chord network, which is the dual structure of the tone network. We could make a comparison with how mechanics deals with the motion of a system of particles, where the motion is decomposed as the motion of the center of mass plus the motion of the particles around the center of mass. In the case of chords, in addition to rotations, we must add the symmetries of a regular polygon, which also includes reflections. The root is the center of a *chord cell*, i.e., the set of chords sharing this note, and plays the role of the center of mass, while the directed chords obtained from symmetric mode intervals applied to the root are equivalent to the system of particles.

In the chord network, neo-Riemannian theory has three primary transformations: the Parallel (P), Relative (R), and Leittonwechsel (L) operations. The P transformation changes a major triad to its parallel minor (and vice versa) by altering the third note (e.g., C to Cm). The R transformation moves a major triad to its relative minor, shifting the fifth up one tone (e.g., C to Am). The L transformation changes a major triad to a minor triad by moving down the root by one semitone (e.g., C to Em). Other transformations must be obtained by combining the above ones. For example, the G minor chord of dominant is obtained by applying PRL on the C major chord of tonic. However, proximity of chords in the chord network is not equivalent to harmonic consonance, although harmonic consonance is somehow correlated with these chord operations. The most efficient voice leading chains between triads of the diatonic scale [2, 34] are obtained from the operations L and R, while the P transformation is less efficient and less consonant.

Figure 1.5 [12] displays the harmonic dissonance of the chords appearing in the tone network of Figure 1.4 around the C major chord by painting darker the chords more dissonant with the central chord.

Two- or three-dimensional geometric models for the chord network allow us to visualize chord transformations, which map out the relationships between chords in a lattice structure. However, higher-dimensional

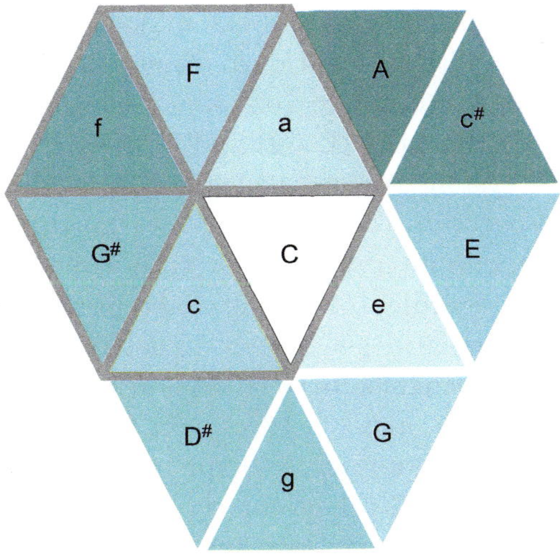

Figure 1.5 Harmonic dissonance of chords in the Tonnetz around the C major chord (minor chords in lowercase). The darker, the more dissonant.

chords cannot be visualized in this way, so that a complete algebra of chords is necessary.

1.7 ALGEBRA OF CHORDS

In the current book, we propose a new and simple formalism to study the structure of chords in an n-TET tuning system and their transformations over the generalized Tonnetz [35] and its dual diagram, the chord network, generated by a given interval structure. In extended just intonation, a mathematical model for navigation through a complex space of harmonies was proposed in [37]. This is an alternative approach to the pitch class set theory for atonal music. It is based on the notion of *mode* used by musicians such as Messiaen [30] and Vieru [36]: an ordered k-tuple of PCs from the n-TET scale defines a directed k-chord (a rolled chord), which is composed of a *root* (the starting note) and a *mode* (a partition of the octave). The rotations of a directed chord define an equivalence class, and the shifts of a mode define a mode class (mode shifts mean rotations of the mode intervals). The notes of a directed chord, as a set where order does not matter, form a chord.

Starting with the particular case of trichords, we describe the structure of the chord cell and the tonal cell, which are mutualy combinatorial dual polytopes .

In addition to the case example of the major and minor triads associated with the mode [3, 4, 5] mentioned above, it is worth mentioning theoretical works using alternative Tonnetze, such as [27] with the mode [0, 1, 3] and [9] with the modes [1, 4, 7] and [2, 3, 7].

For an n-TET system, we introduce and formalize an algebra of chords for a *non-degenerate* Tonnetz, meaning that the generating mode has *non-two equal* mode intervals so that transformations can be done in higher dimensions regardless of geometrical representations. Nevertheless, examples of particular degenerated Tonnetze will also be analyzed.

Transformations distinguish between operations on the root, such as inversion and translation, and operations on the mode, such as mode inversion, retrogradation, and shifts. Some operations are well defined only for directed chords and others for chords, which necessarily are a composition of operations on root and mode.

Translations by mode intervals transform chords within the same mode class, either within the same chord cell or toward another cell. They are equivalent to a combination of mode shifts and chord rotations. However, changing the mode class, although for trichords an inversion suffices, for higher k-dimensional chords must be done through the $k-1$ transpositions of the mode intervals. Among these, a number $k-2$ are associated with one-step *drifts* along the edges within one chord cell, while the remaining drift sends the chord toward a non-congruent chord cell, which generalizes the leading tone exchange of the Neo-Riemannian triadic progressions [10, 11, 18, 26]. In this way it is possible to identify in any dimension simple circuits (closed paths where no vertex is repeated) over the chord cell and over the chord network, such as those that generalize the maximally smooth cycles [10].

This book is an extension of the article [15], which now includes detailed examples and illustrations for trichords and tetrachords to make the formalism more comprehensible, as well as cases that were not studied there.

CHAPTER 2

Modes and chords

2.1 MODE

We will study chords composed of an arbitrary number of PCs in a n-TET scale, which is the tuning system that allows more flexibility to modulate between keys. In a first stage, we shall try to answer some basic questions such as the following ones: how many chords of a number of notes, without counting inversions, are there in a n-TET system? which interval structures produce related sounds? how close these chords are? how many chords have a similar interval structure? how are sequences of chords generated that maintain an invariant part? It is not difficult to answer these and similar questions if we first agree on the terminology.

A k-chord $\{a_0, \ldots, a_{k-1}\}$ is a subset of k different notes out of a total set of n notes, hereafter limited to a n-TET system, therefore equally distributed within the octave. We will represent a note $\frac{k}{n}$ from its ordinal k in a circle of perimeter n, denoted as $\mathcal{O}_n$. In such a case, the notes maintain an easy identifiable interval relationship. We say that the *rank* of the chord is n and the *dimension* is k.

The intrinsic structure of the chord is given by an ordered k-tuple of positive intervals between notes. This quality will be referred to as the *mode* of the chord[1]. The notion of mode as a set of intervals between tones comes from the ancient Greek, generally for defining a type of scale.

The chord is then the result of applying the mode to a particular note. For instance, we may think of applying the mode of a major trichord to

[1]The current approach does not exactly coincide with the view of the PC set theory for atonal music [20]. Therefore, to avoid misunderstandings, we do not use the same concepts of interval vector or interval class vector to describe the interval structure of the chords.

DOI: 10.1201/9781003597964-2

the note C to produce the C major chord. Under this point of view, all the chords of the same mode differ only in a translation.

Usually, a mode is a quality of a non-equally tempered scale, such as the set of consecutive intervals composing the heptatonic major Pythagorean scale, or the seven modes obtained by beginning a major or minor diatonic scales in a different note. When these modes are considered within an equally tempered scale, the mode results in a selection of a subset of notes maintaining a specific interval structure. Hence, the notion of mode is generalized here to any subset of notes of a n-TET system.

The octave $\mathcal{O}_n$ containing the whole set of n notes of the equal temperament scale is represented as points evenly distributed along a circle of n units length, oriented clockwise, and beginning with 0. Numerically, a note of a n-TET system is a number in $\mathbb{Z}_n$.

A k-mode is defined by choosing a sequence of k positive integers, $[A_0, \ldots, A_{k-1}]$, so that their sum is the rank of the chord,

$$A_0 + \cdots + A_{k-1} = n \tag{2.1}$$

Therefore, each positive interval has, at most, $n - k + 1$ unit's length and k is not greater than n. In addition, since an interval of n steps leaves invariant any note in the octave, i.e., it is equivalent to 0 steps, the whole octave $\mathcal{O}_n$ will correspond to the full mode $[0] = [*]$, where the asterisk means the amount left to fill a full octave. We shall assume that the indices of the mode intervals take values in $\mathbb{Z}_k$.

The set of k-modes is notated as $M(n, k)$, where n is the *rank* of the mode and k the *dimension.*

However, we may need to work with modes of rank m, lower than the number of notes of the n-TET system[2]. In order to make such a distinction, and owing to Eq. 2.1, a mode $\mu = [A_0, \ldots, A_{k-1}] \in M(n, k)$ inducing a partition of $\mathcal{O}_n$ into k intervals, i.e., which sum equals n, will be called *full* mode. We can also use the notation

$$\mu = [A_0, \ldots, A_{k-2}, *] \tag{2.2}$$

since one of the intervals is redundant. On the other hand, a mode $\lambda \in M(m, k)$, with $m < n$, will be called *partial* mode. We necessarily write the partial mode without asterisk, as

$$\lambda = [B_0, \ldots, B_{k-1}]; \quad B_0 + \cdots + B_{k-1} = m < n \tag{2.3}$$

[2]Full rank modes are also called successive interval arrays [6] or circular interval series [7], while partial modes are called interval arrays or interval series.

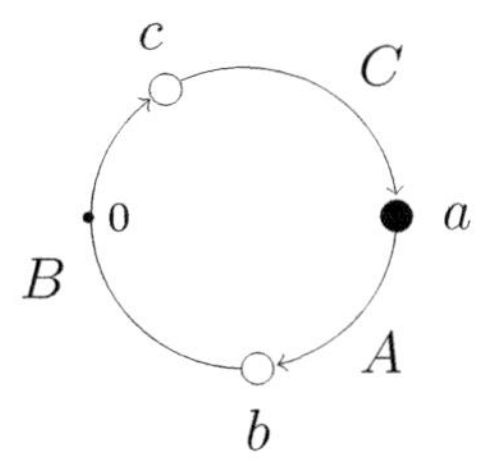

Figure 2.1 The full mode $[A, B, C]$ applied to the note a generates the trichord $\{a, b, c\}$.

Greek characters will be used to represent modes to easily distinguish them from chords and notes.

In addition, in an n-TET system, the notion of mode can also be generalized to a series of interval values giving several turns to the circle of the octave, i.e., with total size larger than n, although, in order to determine the notes of a chord, one must always consider $\bmod n$. Therefore, the dimension k of a *generalized mode* may be larger, equal, or lesser than n.

Thus, a generalized mode may contain intervals from 0 to $n-1$, so that any sequence with components taking values in the set $\mathbb{Z}_n$ of integers modulo n can be interpreted as a set of intervals or as a set of PCs, that is, as a chord. This dual point of view was developed in [36], and its mathematical aspects have been further studied in [1]. However, our aim is not to use this duality but to focus in the interval structure of the chords. Therefore, any interval equal to 0, that is, of full rank, will be omitted in the mode structure, since it does not provide any information about the chord[3].

2.2 DIRECTED CHORD AND CHORD

It is possible to write the same chord by applying several full modes to different notes. We show the example of a trichord, displayed in Figure 2.1. The full mode $[A, B, C]$ is applied to the note a out of a n-TET system, by producing the chord $\{a, b, c\}$.

In general, the same mode applied to different notes produces different chords. However, from Figure 2.1 it is easy to see that the mode

[3]A generalized mode that may contain zeros is referred to as a weak generalized mode.

$[A, B, C]$ applied to a gives the same chord than the mode $[B, C, A]$ applied to b, and then the mode $[C, A, B]$ applied to c, which will be called shifted modes. In this way, we get three *rotations* of the same chord $\{a, b, c\}$. They can be written from the triplets (a, b, c), (b, c, a), and (c, a, b), by maintaining the order clockwise direction on the circle $\mathcal{O}_n$.

Therefore, it is natural to establish an equivalence relation between modes, with one mode class for the three modes $[A, B, C]$, $[B, C, A]$, and $[C, A, B]$; which, in general, are not the same mode (although they are when A, B, C are equal), so that when applied to three different notes they generate the same chord. Similarly, a chord will represent the class of the set of all the permutations of its elements.

The result of applying a full mode $\mu = [A_0, \ldots, A_{k-1}] \in M(n, k)$ to the note $a_0 \in \{0, \ldots, n-1\}$ is called *directed chord*. It is expressed as

$$a \equiv (a_0, \ldots, a_{k-1}) = a_0 \,|\, [A_0, \ldots, A_{k-2}, *] \tag{2.4}$$

with[4]

$$a_1 = a_0 + A_0 \mod n, \;\ldots\;, a_{k-1} = a_{k-2} + A_{k-2} \mod n \tag{2.5}$$

satisfying, in addition, the cyclic property

$$a_0 = a_{k-1} + A_{k-1} \mod n$$

This construction grants the order, clockwise direction around the circle $\mathcal{O}_n$, of the directed chord elements.

However, when a partial mode $\lambda = [B_0, \ldots, B_{k-1}] \in M(m, k)$, with $m < n$, is applied to a note b_0, we obtain a directed chord of dimension $k + 1$,

$$b \equiv (b_0, \ldots, b_{k-1}, b_k) = b_0 \,|\, [B_0, \ldots, B_{k-1}] \tag{2.6}$$

with

$$b_1 = b_0 + B_0 \mod n, \;\ldots\;, b_{k-1} = b_{k-2} + B_{k-2} \mod n$$

and

$$b_0 \neq b_k = b_{k-1} + B_{k-1} \mod n$$

---

[4]The expression $r = p + q \mod n$, although written without parentheses, will mean that r is the reminder in the Euclidean division of $p + q$ by n, otherwise we would write $r = p + (q \mod n)$.

For a directed chord $a \equiv (a_0, \dots, a_{k-1})$, Eq. 2.4 will be written according to the following notation

$$a = a_0 \mid \mu \tag{2.7}$$

being a_0 the *root* of the directed chord[5]. Root and mode are the two components of a directed chord. Hence, given a root a_0 and two full modes $\mu, \nu \in M(n,k)$, $a_0 \mid \mu = a_0 \mid \nu$ if and only if $\mu = \nu$.

From a musical viewpoint, all the directed chords containing the same notes as a, i.e., which are similar to a except in the order of the notes, yield the same chord. Thus, the k-chord must be defined as this equivalence class of directed k-chords, namely

$$\{a\} = \{a_0, \dots, a_{k-1}\} = \{a_0 \mid \mu\} \tag{2.8}$$

which is usually represented by the directed chord $a = (a_0, \dots, a_{k-1})$ whose notes are strictly ordered in the circle of one octave, starting from a_0. Therefore, two directed chords producing the same chord will only differ in a rotation of their notes.

The directed k-chords of rank n are noted as $K(n,k)$, and its classes, the k-chords, are noted as $\mathcal{K}(n,k)$. The set of directed chords belonging to the same chord are the *rotations* of the chord (we also consider the identity or zero rotation). Thus, there are k rotations[6] a^i of a directed k-chord a. In $\mathcal{K}(n,k)$ we operate as it is usual in set theory, by defining subsets, unions, and intersections.

While dealing with directed chords, chords, and modes we shall assume that the *indices* of notes a_i, interval modes A_i, and chord rotations a^i are defined for $i \in \mathbb{Z}_k$, with the sum defined in the group $(\mathbb{Z}_k, +)$.

2.3 REDUCTION OF GENERALIZED MODES

Any generalized mode $\mu \in M(m,k)$, $m \neq n$, can be *reduced* to a full mode $R(\mu) \in M(n,k)$, so that they produce the same chord when they are applied to any root.

It suffices to be true for the root 0, i.e.,

$$\{0 \mid \mu\} = \{0 \mid R(\mu)\}$$

[5]A directed chord can be musically associated with an arpeggio within the octave that begins with the root, with the notes increasing in pitch.

[6]In musical terms, a rotation of the chord is an inversion, which is not the same as an inverted chord. Two chords are inversionally related if the notes composing one chord are the complementary pitch-classes of the other.

Then, for any root a_0, $\{a_0 \,|\, \mu\} = \{a_0 \,|\, R(\mu)\}$. Therefore,

▷ If μ, ν are generalized modes such that applied to a root a_o their chords satisfy $\{a_0 \,|\, \mu\} = \{a_0 \,|\, \nu\}$, then $R(\mu) = R(\nu)$.

2.4 SUBMODES AND SUPERMODES

Any full mode $\mu = [A_0, \ldots, A_{k-1}] \in M(n, k)$ can be decomposed into partial modes of lower rank and dimension by defining a concatenation or composition of modes, such as,

$$\mu = \mu_1 \cdot \mu_2; \quad \mu_1 = [A_0, \ldots, A_{i-1}]; \quad \mu_2 = [A_i, \ldots, A_{k-1}]; \quad 1 \leq i \leq k-1 \tag{2.9}$$

so that $\text{rank}(\mu_1) + \text{rank}(\mu_2) = n$, and $\dim(\mu_1) + \dim(\mu_2) = k$. Both partial modes μ_1 and μ_2 are then *complementary* in $M(n, k)$.

Any full mode μ can be totally decomposed into $k-1$ concatenations of single modes, $\mu = [A_0] \ldots [A_{k-1}]$, whose intervals equal n (the dot for the composition will be usually omitted).

Conversely, if $\mu_1 = [B_0, \ldots, B_{k_1-1}]$ and $\mu_2 = [C_0, \ldots, C_{k_2-1}]$ are generalized modes, the composition

$$\mu_1 \cdot \mu_2 = [B_0, \ldots, B_{k_1-1}][C_0, \ldots, C_{k_2-1}] \tag{2.10}$$

is defined as the mode that produces the partition of the octave resulting from the successive application of the intervals of μ_1 and μ_2. In particular, this is also valid for full modes.

The composition of generalized modes has a monoid structure, i.e., is associative and has an identity element, which is $[0]$.

In general, if μ_1 and μ_2 are generalized modes, $\mu_1 \cdot \mu_2 \neq R(\mu_1) \cdot R(\mu_2)$ and $R(\mu_1 \cdot \mu_2) \neq R(R(\mu_1) \cdot R(\mu_2))$. However, full modes with the operation $\mu_1 \odot \mu_2 \equiv R(\mu_1 \cdot \mu_2)$ are a commutative group.

We may also define a non-commutative sum of generalized modes as follows,

$$\mu_1 + \mu_2 = [B_0, \ldots, B_{k_1-1} + C_0, \ldots, C_{k_2-1}] \tag{2.11}$$

which may result in a generalized mode.

The sum has also a monoid structure with identity element $[0]$.

In general, if μ_1 and μ_2 are generalized modes, $\mu_1 + \mu_2 \neq R(\mu_1) + R(\mu_2)$ and $R(\mu_1 + \mu_2) \neq R(R(\mu_1) + R(\mu_2))$. However, full modes with the operation $\mu_1 \oplus \mu_2 \equiv R(\mu_1 + \mu_2)$ are a commutative group.

▷ The composition and the sum of *full* modes satisfy

$$R(\mu_1 \cdot \mu_2) = R(\mu_1 + \mu_2) \tag{2.12}$$

We prove it. When applying the generalized modes $\mu_1 \cdot \mu_2$ and $\mu_1 + \mu_2$ to the root 0, they describe the following similar chords,

$$\{0 \,|\, \mu_1 \cdot \mu_2\} = \{0 \,|\, [A_0, A_1, \dots, A_{i-2}, A_{i-1}] \cdot [B_0, B_1, \dots, B_{j-2}, B_{j-1}]\} =$$

$$= \{0, A_0, A_0 + A_1, \dots, A_0 + \dots + A_{i-2}, 0, B_0, \dots, B_0 + \dots + B_{j-2}, 0\}$$

$$\{0 \,|\, \mu_1 + \mu_2\} = \{0 \,|\, [A_0, A_1, \dots, A_{i-2}, A_{i-1} + B_0, B_1, \dots, B_{j-2}, B_{j-1}]\} =$$

$$= \{0, A_0, A_0 + A_1, \dots, A_0 + \dots + A_{i-2}, B_0, \dots, B_0 + \dots + B_{j-2}, 0\}$$

Hence, the respective modes have the same reduced form.

This property has its counterpart for chords.

▷ If μ_1, μ_2 are full modes, the chord $\{a_0 \,|\, \mu_1 \cdot \mu_2\}$ satisfies

$$\{a_0 \,|\, \mu_1 \cdot \mu_2\} = \{a_0 \,|\, \mu_1 + \mu_2\} = \{a_0 \,|\, \mu_1\} \cup \{a_0 \,|\, \mu_2\} \tag{2.13}$$

Both operations allow us to break a full mode in parts in order to study subchords of a given chord.

▷ The *submodes* of a full mode $\mu = [A_0] \cdots [A_{k-1}] \in M(n,k)$ are obtained as $\lambda = [A_0] \circ \cdots \circ [A_{k-1}]$, where each symbol $\circ$ means either a composition or a sum.

In this way we will obtain a number of full modes of the same rank but of different dimension, which may vary from 1 to k.

When both modes μ and λ are applied to the same root a_0, the chord $\{a_0 \,|\, \lambda\}$ gives a partition of the octave less fine than the chord $\{a_0 \,|\, \mu\}$. Hence, we write $\lambda \prec \mu$. Conversely, we say that μ is a *supermode* of λ, and write $\mu \succ \lambda$.

2.5 COMPLEMENTARY MODES

For any rank $m < n$, a partial mode $\mu = [A_0, \dots, A_{p-2}] \in M(m, p-1)$ can be *completed* on the right-hand side up to a full mode in $M(n,p)$ as

$$[\mu, *] \equiv [A_0, \dots, A_{p-2}, n - m] = [A_0, \dots, A_{p-2}, *] \tag{2.14}$$

We express this mode as above or, alternatively, as a concatenation of partial modes

$$[\mu, *] = \mu \cdot [n - m] \tag{2.15}$$

Similarly, the partial mode can be completed on the left-hand side as

$$[*, \mu] = [n - m] \cdot \mu \tag{2.16}$$

Any directed chord $a = a_0 \,|\, [A_0, \ldots, A_{p-2}]$ obtained from a partial mode can obviously be written as $a = a_0 \,|\, [A_0, \ldots, A_{p-2}, *]$, in terms of the completed mode to the right. Therefore, the directed chord a may be written in either of the following forms

$$a = a_0 \,|\, \mu = a_0 \,|\, [\mu, *] \tag{2.17}$$

Any full mode can be written as the concatenation of two *complementary* partial modes, $\mu \cdot \kappa \in M(n, p)$, with $\mu \in M(m, q)$, $n > m$ and $p > q$, and $\kappa \in M(n-m, p-q)$. Hence, the composition of the completed modes satisfies

$$\mu \cdot \kappa = [\mu, *] \cdot [*, \kappa] \tag{2.18}$$

allowing us to write the composition of complementary modes as a composition of full modes.

Similarly, the sum of completed modes satisfies

$$[\mu, *] + [*, \kappa] = [\mu, *] \cdot [*, \kappa] \tag{2.19}$$

Therefore, Eq. 2.19 becomes a particular case of Eq. 2.12.

On the other hand, we may assume that a full mode λ is equivalent to the completed modes $[\lambda, *]$ and $[*, \lambda]$, and that any couple of full modes are complementary. This allows to interpret Eq. 2.12 as a particular case of Eq. 2.19. In other words, Eq. 2.19 is valid for complementary partial modes, as well as for full modes.

For a chord $\{a_0 \,|\, \mu \cdot \kappa\} \in \mathcal{K}(n, p)$ generated by the composition of two complementary modes, by taking into account Eq. 2.18, we obtain a particular case of Eq. 2.13

$$\{a_0 \,|\, \mu \cdot \kappa\} = \{a_0 \,|\, [\mu, *]\} \cup \{a_0 \,|\, [*, \kappa]\} \tag{2.20}$$

Obviously, the modes $[\mu, *]$ and $[*, \kappa]$ are submodes of $\mu \cdot \kappa$, and the chords $\{a_0 \,|\, [\mu, *]\}$ and $\{a_0 \,|\, [*, \kappa]\}$ are subchords of $\{a_0 \,|\, \mu \cdot \kappa\}$.

2.6 MODE SHIFTS AND DIRECTED CHORD ROTATIONS

The equivalence relation given by the rotations of the directed chords may be transferred to the modes. Following the example in the beginning of Section 2.2, the rotations of a directed chord are generated by applying the i-th shifted full mode to the i-th note of the chord, as defined below. Therefore, we shall introduce the *mode class* in regard to the shifts of a mode.

For a full mode $\mu \equiv \mu^0 = [A_0, \ldots, A_{k-1}] \in M(n,k)$ we define a single *shift* as an one-step cyclic permutation of the mode intervals, such as

$$\mu^1 = [A_1, \ldots, A_{k-1}, A_0]$$

A m-shift (shift of m steps) of μ, with $m \in \mathbb{Z}_k$, is defined as

$$\mu^m = [A_m, A_{m+1}, \ldots, A_{m+k-1}] = [A_m, A_{m+1}, \ldots, A_{m-1}] \tag{2.21}$$

For $m = k$ the mode becomes the original mode or the 0-shift of μ. Al the shifts of a mode form a mode class.

Notice that a single shift is obtained by the permutation $\sigma = (k, 1, \ldots, k-1)$ of the symmetric group $\mathcal{S}_k$, with $k!$ elements. Since $\sigma^k = 1$, a shift generates a cyclic subgroup $\langle\sigma\rangle$ of $\mathcal{S}_k$ of order k, which provides a number $(k-1)!$ of cosets in $\mathcal{S}_k/\langle\sigma\rangle$. The group $\mathcal{S}_k$ may be obtained from generators in several ways, e.g., from the product of transpositions in the form $(1, i)$, $1 < i \leq k$, or from products of σ and the transposition $(1, 2)$.

Thus, we are able to obtain k equivalent rotations of a directed chord $a = (a_0, \ldots, a_{k-1})$ by applying the elements of a mode class to k different roots, as follows

$$a^m = a_m \mid \mu^m = (a_m, \ldots, a_{m+k-1}) = (a_m, \ldots, a_{m-1}) \tag{2.22}$$

with all the indices in $\mathbb{Z}_k$.

Notice that, depending on the interval shape, it is possible to get less than k different shifted modes. For example, if $A_0 = \cdots = A_{k-1}$, all the shifts produce the same mode, although there are still k rotations of any directed chord. Therefore, the number of rotations is associated with the k different roots where the shifted modes are applied, and not with the number of elements of the mode class.

The set of mode classes of $M(n,k)$ with regard to the shifts, for all the possible interval values satisfying $A_0 + A_1 + \cdots + A_{k-1} = n$, is notated as $\mathcal{M}(n,k)$. The set of mode classes for the intervals of a mode

$\mu \in M(n,k)$ is notated as $\mathcal{C}(\mu)$. The mode class of μ, namely $\mu^S \in \mathcal{C}(\mu)$, is composed of the non-equal shifts μ^m, for $m \in \{0,\ldots,k-1\}$, i.e., is a *fixed necklace* of length k. The number of elements composing a mode class μ^S is notated as $s(\mu^S)$, which is equivalent to the number $s(\mu)$ of different shifts of any mode of the class.

If the mode intervals are non-two equal, then the notes composing a directed chord $a_0 \,|\, [A_0,\ldots,A_{k-1}]$ are univocally determined. In addition, for a non-degenerate Tonnetz the mode classes μ^S do not match, i.e., $s(\mu) = k$, and $\mathcal{C}(\mu) = (k-1)!$. If the mode intervals are coprime, the number of different chords in the chord network is $n(k-1)!$ and contain all the notes of the n-TET scale.

For example, three non-equal intervals satisfying $A + B + C = n$ generate two mode classes. One is composed of the mode $\mu = [A,B,C]$ together with its shifts μ^1, μ^2, that we notate as μ^S, with a number of shifts $s(\mu) = 3$. The other mode class is composed of the modes $\nu = [A,C,B]$ and its shifts ν^1, ν^2, written as ν^S, with a number of shifts $s(\nu) = 3$. Thus, we write $\mathcal{C}(\mu) = \mathcal{C}(\nu) = \{\mu^S, \nu^S\}$.

2.7 NUMBER OF CHORDS AND MODES

Let us review some basic concepts of combinatorics. In a set A of n elements, the number of ways of choosing a subset $B \subset A$ of k elements is given by the number

$$C(n,k) \equiv \binom{n}{k} = \frac{n!}{k!(n-k)!}$$

which is referred to as k-combinations of a n-set.

As the empty subset ϕ is also considered, the total number of subsets that can be obtained with the elements of A is $\sum_{k=0}^{n} \binom{n}{k} = 2^n$.

An example of combinations is how to compute the number of *weak compositions* of a positive integer p into q parts ($q \geq 1$), that is, the way of writing p as the sum of a sequence of q non-negative integers. This means:

$$x_1 + \cdots + x_q = p$$

so that p units are distributed in q boxes (variables) that are separated by $q-1$ dividers (+ signs). For example, if $p = 9$ and $q = 3$, one possible distribution is

$$\circ\circ\circ \,|\, \circ\circ \,|\, \circ\circ\circ\circ$$

Hence, in general, we must choose $q-1$ dividers out of $p+q-1$ items, either units or dividers. This provides a number $\binom{p+q-1}{q-1}$ of weak compositions of p elements into k parts. Notice that it is possible a combination in the form

$$\circ\ \circ\ \circ \mid \mid \circ\ \circ\ \circ\ \circ\ \circ\ \circ$$

meaning that one or more variables may be zero.

Similarly, the number of *compositions* of a positive integer p into q parts is the way of writing p as the sum of a sequence of q ($q \leq p$) strictly positive integers. In this case, there must be at least one unit in each box. Therefore, there are $p-q$ units left, which have to be distributed into q boxes as in the case of weak compositions. Thus, the total number of items to distribute is $(p-q)+q-1$, out of which a number $q-1$ are dividers. This leads to a number

$$C(p-1, q-1) = \binom{p-1}{q-1}$$

of compositions of p elements into q parts.

The total number of compositions of a p-set is $\sum_{q=1}^{p} \binom{p-1}{q-1} = 2^{p-1}$.

A k-chord has been defined as a subset of k notes out of a n notes system, regardless the order. Hence, the number of k-chords in $\mathcal{K}(n,k)$ is given straightforward by the k-combinations of n elements,

$$N_{\mathcal{K}}(n,k) = C(n,k) \tag{2.23}$$

The number of subchords of dimensions from 1 to n in $\mathcal{K}(n,k)$ is, obviously, $2^n - 1$.

For instance, the number of trichords is:

- in a 12-TET system, $N_{\mathcal{K}}(12,3) = \binom{12}{3} = 220$
- in a 53-TET system, $N_{\mathcal{K}}(53,3) = \binom{53}{3} = 23,426$

Notice that the number of chords does not depend on whether the scale is of equal temperament. However, equal temperament is a necessary condition to work with modes; not in order to define the shifts, but to grant that every shift applied to any note of the scale produces a chord with notes of the scale.

On the other hand, the case of compositions is just the situation of evaluating the number of modes in $M(n,k)$. Hence, the number of modes is

$$N_M(n,k) = C(n-1,k-1) \tag{2.24}$$

Thus, the number of trimodes is:

- in a 12-TET system, $N_M(12,3) = \binom{12-1}{3-1} = 55$
- in a 53-TET system, $N_M(53,3) = \binom{53-1}{3-1} = 1,324$

In regard to submodes, if $\lambda = [C_1, \ldots, C_m] \in M(n,m)$ is a submode of $\mu = [B_1, \ldots, B_k] \in M(n,k)$, each interval C_i is the sum of several consecutive intervals B_j, so that each interval B_j is taken into account only once. Then, the number of submodes $\lambda \in M(n,m)$ of $\mu \in M(n,k)$, $1 \leq m \leq k$, is the number of compositions of k elements into m parts, that is, $C(k-1, m-1)$. The total number of submodes of μ up to dimension k is then 2^{k-1}, by including the k-mode μ and the 1-mode $[0]$.

For example, in any n-TET system, the number of submodes of any 3-mode is $2^{3-1} = 4$, from which $C(3-1,1-1) = 1$ is the 1-mode $[0]$, $C(3-1,2-1) = 3$ are 2-modes, and $C(3-1,3-1) = 1$ is the initial 3-mode.

On the other hand, in the 53-TET scale, since $n \neq \dot{3}$, there is no 3-mode in the form $[A_0, A_0, A_0]$. Hence, all mode classes have 3 modes, and the number of mode classes is:

- in a 53-TET system, $N_{\mathcal{M}}(53,3) = 1,324/3 = 442$

The full sets of 3-modes and mode classes for a 53-TET system are listed in Table 2.3, according to the same previous notation.

2.8 NUMBER OF MODE CLASSES

The number of mode classes in $\mathcal{M}(n,k)$, namely $N_{\mathcal{M}}(n,k)$, depends on the values n and k, since it depends on the number of k-modes that remain invariant by shifts. The way of computing such a number of mode classes will be shown with two examples of trimodes for the 12-TET and 53-TET systems.

In a 12-TET scale, the 55 trimodes have one of the following shapes:

- modes with different intervals, say $[A_0, A_1, A_2]$, that have three possible shifts, namely, $[A_0, A_1, A_2]$, $[A_1, A_2, A_0]$, and $[A_2, A_0, A_1]$

Table 2.1 List of 55 3-modes comprising 19 mode classes in regard to shifts for 12-TET.

1	1	10	1	1	2	2	8	2	2	3	3	6	3	3	4	4	4
1	2	9	1	2	2	3	7	2	3	3	4	5	3	4			
1	3	8	1	3	2	4	6	2	4	3	5	4	3	5			
1	4	7	1	4	2	5	5	2	5								
1	5	6	1	5	2	6	4	2	6								
1	6	5	1	6	2	7	3	2	7								
1	7	4	1	7													
1	8	3	1	8													
1	9	2	1	9													

- modes with 2 similar intervals, say $[A_0, A_0, A_1]$, that also have the three different shifts $[A_0, A_0, A_1]$, $[A_0, A_1, A_0]$, and $[A_1, A_0, A_0]$
- one mode in the form $[A_0, A_0, A_0]$, since in this case $n = \dot{3}$. It is, therefore, invariant under shifts

Therefore, the number of 3-mode classes is:

- in a 12-TET system, $N_{\mathcal{M}}(12, 3) = \frac{55-1}{3} + 1 = 19$

In Table 2.1 are listed the 3-mode classes (packs of consecutive numbers on white background) for the 12-TET system, as well as the full set of modes when they are read by beginning with a number on white followed by the 2 subsequent numbers, either on white or gray background. Therefore, there are as many different 3-modes as numbers on white background. Each class is composed of the mode on white background in addition to the shifts ending in the numbers on the gray.

Let us see another example: the hexachords of the 12-TET system, listed in Table 2.2. In this case, the number of 6-chords and 6-modes is $N_{\mathcal{K}}(12, 6) = 924$ and $N_{M}(12, 6) = 462$, respectively. Most of the modes are grouped in equivalence classes with 6 shifts, but there are some belonging to classes with less shifts. The modes with a periodic pattern $[A_0, A_1, A_2, A_1, A_2, A_3]$ (we assume the intervals are not the same one), $[A_0, A_1, A_0, A_1, A_0, A_1]$, and $[A_0, A_0, A_0, A_0, A_0, A_0]$ will belong to mode classes with 3, 2, and 1 shifts, respectively. In particular:

- For the modes in the form $[A_0, A_1, A_2, A_1, A_2, A_3]$ their intervals satisfy $A_0 + A_1 + A_2 = 6$. Hence the intervals can take the set of values $1, 1, 4$ or $1, 2, 3$. Thus, the modes $[1, 1, 4, 1, 1, 4]$, $[1, 2, 3, 1, 2, 3]$, and $[1, 3, 2, 1, 3, 2]$ belong to 3 different mode classes with 3 shifts.

Table 2.2 List of 462 6-modes comprising 80 mode classes in regard to shifts for 12-TET.

1	1	1	1	1	7	1	1	1	1	1	1	1	1	6	1	2	1	1	1	6	1	1	1	3	2	1	4	1	1	3	2	1	1	2	1	3	3	2	1	2	1	3	3
1	1	1	1	2	6	1	1	1	1	2	1	1	2	1	1	6	1	1	2	1	1	1	1	3	2	2	3	1	1	3	2	2	1	2	1	4	1	3	1	2	1	4	1
1	1	1	1	3	5	1	1	1	1	3	1	1	2	1	2	5	1	1	2	1	2	1	1	3	2	3	2	1	1	3	2	3	1	2	1	4	2	2	1	2	1	4	2
1	1	1	1	4	4	1	1	1	1	4	1	1	2	1	3	4	1	1	2	1	3	1	1	3	3	1	3	1	1	3	3	1	1	2	2	1	2	4	1	2	2	1	2
1	1	1	1	5	3	1	1	1	1	5	1	1	2	1	4	3	1	1	2	1	4	1	1	3	3	2	2	1	1	3	3	2	1	2	2	1	3	3	1	2	2	1	3
1	1	1	1	6	2	1	1	1	1	6	1	1	2	1	5	2	1	1	2	1	5	1	1	3	4	1	2	1	1	3	4	1	1	2	2	1	4	2	1	2	2	1	4
1	1	1	2	1	6	1	1	1	2	1	1	1	2	2	1	5	1	1	2	2	1	1	1	4	1	1	4	1	1				1	2	2	2	1	4	1	2	2	2	1
1	1	1	2	2	5	1	1	1	2	2	1	1	2	2	2	4	1	1	2	2	2	1	1	4	1	2	3	1	1	4	1	2	1	2	2	2	2	3	1	2	2	2	2
1	1	1	2	3	4	1	1	1	2	3	1	1	2	2	3	3	1	1	2	2	3	1	1	4	1	3	2	1	1	4	1	3	1	2	2	2	3	2	1	2	2	2	3
1	1	1	2	4	3	1	1	1	2	4	1	1	2	2	4	2	1	1	2	2	4	1	1	4	2	1	3	1	1	4	2	1	1	2	2	3	1	3	1	2	2	3	1
1	1	1	2	5	2	1	1	1	2	5	1	1	2	3	1	4	1	1	2	3	1	1	1	4	2	2	2	1	1	4	2	2	1	2	2	3	2	2	1	2	2	3	2
1	1	1	3	1	5	1	1	1	3	1	1	1	2	3	2	3	1	1	2	3	2	1	1	4	3	1	2	1	1	4	3	1	1	2	3	1	2	3	1	2			
1	1	1	3	2	4	1	1	1	3	2	1	1	2	3	3	2	1	1	2	3	3	1	1	5	1	2	2	1	1	5	1	2	1	2	3	1	3	2	1	2	3	1	3
1	1	1	3	3	3	1	1	1	3	3	1	1	2	4	1	3	1	1	2	4	1	1	1	5	2	1	2	1	1	5	2	1	1	2	3	2	1	3	1	2	3	2	1
1	1	1	3	4	2	1	1	1	3	4	1	1	2	4	2	2	1	1	2	4	2	1	2	1	2	1	5	1	2	1	2	1	1	2	3	2	2	2	1	2	3	2	2
1	1	1	4	1	4	1	1	1	4	1	1	1	2	5	1	2	1	1	2	5	1	1	2	1	2	2	4	1	2	1	2	2	1	3	1	3	1	3	1				
1	1	1	4	2	3	1	1	1	4	2	1	1	3	1	1	5	1	1	3	1	1	1	2	1	2	3	3	1	2	1	2	3	1	3	1	3	2	2	1	3	1	3	2
1	1	1	4	3	2	1	1	1	4	3	1	1	3	1	2	4	1	1	3	1	2	1	2	1	2	4	2	1	2	1	2	4	1	3	2	1	3	2	1	3			
1	1	1	5	1	3	1	1	1	5	1	1	1	3	1	3	3	1	1	3	1	3	1	2	1	3	1	4	1	2	1	3	1	1	3	2	2	2	2	1	3	2	2	2
1	1	1	5	2	2	1	1	1	5	2	1	1	3	1	4	2	1	1	3	1	4	1	2	1	3	2	3	1	2	1	3	2	2	2	2	2	2	2					

Table 2.3 List of 1,324 3-modes comprising 442 3-mode classes in regard to shifts for 53-TET.

1 1 51	1 1	2 2 49	2 2	3 3 47	3 3	4 4 45	4 4	5 5 43	5 5	6 6 41	6 6	7 7 39	7 7	8 8 37	8 8	9 9 35	9 9	10 10 33	10 10	11 11 31	11 11	12 12 29	12 12	13 13 27	13 13	14 14 25	14 14	15 15 23	15 15	16 16 21	16 16	17 17 19	17 17
1 2 50	1 2	2 3 48	2 3	3 4 46	3 4	4 5 44	4 5	5 6 42	5 6	6 7 40	6 7	7 8 38	7 8	8 9 36	8 9	9 10 34	9 10	10 11 32	10 11	11 12 30	11 12	12 13 28	12 13	13 14 26	13 14	14 15 24	14 15	15 16 22	15 16	16 17 20	16 17	17 18 18	17 18
1 3 49	1 3	2 4 47	2 4	3 5 45	3 5	4 6 43	4 6	5 7 41	5 7	6 8 39	6 8	7 9 37	7 9	8 10 35	8 10	9 11 33	9 11	10 12 31	10 12	11 13 29	11 13	12 14 27	12 14	13 15 25	13 15	14 16 23	14 16	15 17 21	15 17	16 18 19	16 18		
1 4 48	1 4	2 5 46	2 5	3 6 44	3 6	4 7 42	4 7	5 8 40	5 8	6 9 38	6 9	7 10 36	7 10	8 11 34	8 11	9 12 32	9 12	10 13 30	10 13	11 14 28	11 14	12 15 26	12 15	13 16 24	13 16	14 17 22	14 17	15 18 20	15 18	16 19 18	16 19		
1 5 47	1 5	2 6 45	2 6	3 7 43	3 7	4 8 41	4 8	5 9 39	5 9	6 10 37	6 10	7 11 35	7 11	8 12 33	8 12	9 13 31	9 13	10 14 29	10 14	11 15 27	11 15	12 16 25	12 16	13 17 23	13 17	14 18 21	14 18	15 19 19	15 19	16 20 17	16 20		
1 6 46	1 6	2 7 44	2 7	3 8 42	3 8	4 9 40	4 9	5 10 38	5 10	6 11 36	6 11	7 12 34	7 12	8 13 32	8 13	9 14 30	9 14	10 15 28	10 15	11 16 26	11 16	12 17 24	12 17	13 18 22	13 18	14 19 20	14 19	15 20 18	15 20				
1 7 45	1 7	2 8 43	2 8	3 9 41	3 9	4 10 39	4 10	5 11 37	5 11	6 12 35	6 12	7 13 33	7 13	8 14 31	8 14	9 15 29	9 15	10 16 27	10 16	11 17 25	11 17	12 18 23	12 18	13 19 21	13 19	14 20 19	14 20	15 21 17	15 21				
1 8 44	1 8	2 9 42	2 9	3 10 40	3 10	4 11 38	4 11	5 12 36	5 12	6 13 34	6 13	7 14 32	7 14	8 15 30	8 15	9 16 28	9 16	10 17 26	10 17	11 18 24	11 18	12 19 22	12 19	13 20 20	13 20	14 21 18	14 21	15 22 16	15 22				
1 9 43	1 9	2 10 41	2 10	3 11 39	3 11	4 12 37	4 12	5 13 35	5 13	6 14 33	6 14	7 15 31	7 15	8 16 29	8 16	9 17 27	9 17	10 18 25	10 18	11 19 23	11 19	12 20 21	12 20	13 21 19	13 21	14 22 17	14 22						
1 10 42	1 10	2 11 40	2 11	3 12 38	3 12	4 13 36	4 13	5 14 34	5 14	6 15 32	6 15	7 16 30	7 16	8 17 28	8 17	9 18 26	9 18	10 19 24	10 19	11 20 22	11 20	12 21 20	12 21	13 22 18	13 22	14 23 16	14 23						
1 11 41	1 11	2 12 39	2 12	3 13 37	3 13	4 14 35	4 14	5 15 33	5 15	6 16 31	6 16	7 17 29	7 17	8 18 27	8 18	9 19 25	9 19	10 20 23	10 20	11 21 21	11 21	12 22 19	12 22	13 23 17	13 23	14 24 15	14 24						
1 12 40	1 12	2 13 38	2 13	3 14 36	3 14	4 15 34	4 15	5 16 32	5 16	6 17 30	6 17	7 18 28	7 18	8 19 26	8 19	9 20 24	9 20	10 21 22	10 21	11 22 20	11 22	12 23 18	12 23	13 24 16	13 24								
1 13 39	1 13	2 14 37	2 14	3 15 35	3 15	4 16 33	4 16	5 17 31	5 17	6 18 29	6 18	7 19 27	7 19	8 20 25	8 20	9 21 23	9 21	10 22 21	10 22	11 23 19	11 23	12 24 17	12 24	13 25 15	13 25								
1 14 38	1 14	2 15 36	2 15	3 16 34	3 16	4 17 32	4 17	5 18 30	5 18	6 19 28	6 19	7 20 26	7 20	8 21 24	8 21	9 22 22	9 22	10 23 20	10 23	11 24 18	11 24	12 25 16	12 25	13 26 14	13 26								
1 15 37	1 15	2 16 35	2 16	3 17 33	3 17	4 18 31	4 18	5 19 29	5 19	6 20 27	6 20	7 21 25	7 21	8 22 23	8 22	9 23 21	9 23	10 24 19	10 24	11 25 17	11 25	12 26 15	12 26										
1 16 36	1 16	2 17 34	2 17	3 18 32	3 18	4 19 30	4 19	5 20 28	5 20	6 21 26	6 21	7 22 24	7 22	8 23 22	8 23	9 24 20	9 24	10 25 18	10 25	11 26 16	11 26	12 27 14	12 27										
1 17 35	1 17	2 18 33	2 18	3 19 31	3 19	4 20 29	4 20	5 21 27	5 21	6 22 25	6 22	7 23 23	7 23	8 24 21	8 24	9 25 19	9 25	10 26 17	10 26	11 27 15	11 27	12 28 13	12 28										
1 18 34	1 18	2 19 32	2 19	3 20 30	3 20	4 21 28	4 21	5 22 26	5 22	6 23 24	6 23	7 24 22	7 24	8 25 20	8 25	9 26 18	9 26	10 27 16	10 27	11 28 14	11 28												
1 19 33	1 19	2 20 31	2 20	3 21 29	3 21	4 22 27	4 22	5 23 25	5 23	6 24 23	6 24	7 25 21	7 25	8 26 19	8 26	9 27 17	9 27	10 28 15	10 28	11 29 13	11 29												
1 20 32	1 20	2 21 30	2 21	3 22 28	3 22	4 23 26	4 23	5 24 24	5 24	6 25 22	6 25	7 26 20	7 26	8 27 18	8 27	9 28 16	9 28	10 29 14	10 29	11 30 12	11 30												
1 21 31	1 21	2 22 29	2 22	3 23 27	3 23	4 24 25	4 24	5 25 23	5 25	6 26 21	6 26	7 27 19	7 27	8 28 17	8 28	9 29 15	9 29	10 30 13	10 30														
1 22 30	1 22	2 23 28	2 23	3 24 26	3 24	4 25 24	4 25	5 26 22	5 26	6 27 20	6 27	7 28 18	7 28	8 29 16	8 29	9 30 14	9 30	10 31 12	10 31														
1 23 29	1 23	2 24 27	2 24	3 25 25	3 25	4 26 23	4 26	5 27 21	5 27	6 28 19	6 28	7 29 17	7 29	8 30 15	8 30	9 31 13	9 31	10 32 11	10 32														
1 24 28	1 24	2 25 26	2 25	3 26 24	3 26	4 27 22	4 27	5 28 20	5 28	6 29 18	6 29	7 30 16	7 30	8 31 14	8 31	9 32 12	9 32																
1 25 27	1 25	2 26 25	2 26	3 27 23	3 27	4 28 21	4 28	5 29 19	5 29	6 30 17	6 30	7 31 15	7 31	8 32 13	8 32	9 33 11	9 33																
1 26 26	1 26	2 27 24	2 27	3 28 22	3 28	4 29 20	4 29	5 30 18	5 30	6 31 16	6 31	7 32 14	7 32	8 33 12	8 33	9 34 10	9 34																
1 27 25	1 27	2 28 23	2 28	3 29 21	3 29	4 30 19	4 30	5 31 17	5 31	6 32 15	6 32	7 33 13	7 33	8 34 11	8 34																		
1 28 24	1 28	2 29 22	2 29	3 30 20	3 30	4 31 18	4 31	5 32 16	5 32	6 33 14	6 33	7 34 12	7 34	8 35 10	8 35																		
1 29 23	1 29	2 30 21	2 30	3 31 19	3 31	4 32 17	4 32	5 33 15	5 33	6 34 13	6 34	7 35 11	7 35	8 36 9	8 36																		
1 30 22	1 30	2 31 20	2 31	3 32 18	3 32	4 33 16	4 33	5 34 14	5 34	6 35 12	6 35	7 36 10	7 36																				
1 31 21	1 31	2 32 19	2 32	3 33 17	3 33	4 34 15	4 34	5 35 13	5 35	6 36 11	6 36	7 37 9	7 37																				
1 32 20	1 32	2 33 18	2 33	3 34 16	3 34	4 35 14	4 35	5 36 12	5 36	6 37 10	6 37	7 38 8	7 38																				
1 33 19	1 33	2 34 17	2 34	3 35 15	3 35	4 36 13	4 36	5 37 11	5 37	6 38 9	6 38																						
1 34 18	1 34	2 35 16	2 35	3 36 14	3 36	4 37 12	4 37	5 38 10	5 38	6 39 8	6 39																						
1 35 17	1 35	2 36 15	2 36	3 37 13	3 37	4 38 11	4 38	5 39 9	5 39	6 40 7	6 40																						
1 36 16	1 36	2 37 14	2 37	3 38 12	3 38	4 39 10	4 39	5 40 8	5 40																								
1 37 15	1 37	2 38 13	2 38	3 39 11	3 39	4 40 9	4 40	5 41 7	5 41																								
1 38 14	1 38	2 39 12	2 39	3 40 10	3 40	4 41 8	4 41	5 42 6	5 42																								
1 39 13	1 39	2 40 11	2 40	3 41 9	3 41	4 42 7	4 42																										
1 40 12	1 40	2 41 10	2 41	3 42 8	3 42	4 43 6	4 43																										
1 41 11	1 41	2 42 9	2 42	3 43 7	3 43	4 44 5	4 44																										
1 42 10	1 42	2 43 8	2 43	3 44 6	3 44																												
1 43 9	1 43	2 44 7	2 44	3 45 5	3 45																												
1 44 8	1 44	2 45 6	2 45	3 46 4	3 46																												
1 45 7	1 45	2 46 5	2 46																														
1 46 6	1 46	2 47 4	2 47																														
1 47 5	1 47	2 48 3	2 48																														
1 48 4	1 48																																
1 49 3	1 49																																
1 50 2	1 50																																

- For the modes in the form $[A_0, A_1, A_0, A_1, A_0, A_1]$ their intervals satisfy $A_0 + A_1 = 4$. Hence the intervals can take the values 1 or 3. Then, the mode $[1, 3, 1, 3, 1, 3]$ belongs to one mode class with 2 shifts.

- The mode $[A_0, A_0, A_0, A_0, A_0, A_0]$ is only possible for $A_0 = 2$. The mode $[2, 2, 2, 2, 2, 2]$ belongs to one mode class with a single shift.

Therefore, there are 12 modes comprising the 5 mode classes with less than 6 shifts. The number of 6-mode classes is,

- in a 12-TET system, $N_{\mathcal{M}}(12, 6) = \frac{462-12}{6} + 5 = 80$

2.9 COUNTING CHORDS AND MODES

Every k-mode applied to any note of the n-TET system produces a directed chord, clockwise direction around the circle $\mathcal{O}_n$ of the octave. Since each directed k-chord has k rotations, the total number of k-chords is

$$N_{\mathcal{K}}(n, k) = \frac{n}{k} N_M(n, k) \tag{2.25}$$

The number of k-modes $N_M(n, k) = C(n-1, k-1)$ may be also computed in terms of the mode classes by taking into account the number of different shifts of their elements. Thus,

$$N_M(n, k) = \sum_{\lambda^S \in \mathcal{M}(n,k)} s(\lambda^S) \tag{2.26}$$

We have shown how any k-chord can be obtained from k rotations of a directed chord. In other words, the k-chord can be generated by applying the shifted modes to k different root notes. But, for the same reason, it is also possible to produce any chord by applying one particular mode to the appropriate root, so that other shifted modes of the same mode class, applied according to Eq. 2.22, will only produce rotations of the same chord. Therefore, the number of k-chords in an n-TET system accounts only for the set of mode classes $\mathcal{M}(n, k)$, one mode for each class.

Thus, for a mode class $\lambda^S \in \mathcal{M}(n, k)$ with $s(\lambda^S)$ elements, i.e., the number of different shifts, since every k-chord has k rotations, we may evaluate the number of k-chords generated by λ^S as $\frac{n\,s(\lambda^S)}{k}$, so that the total number of k-chords is computed as

$$N_{\mathcal{K}}(n,k) = \sum_{\lambda^S \in \mathcal{M}(n,k)} \frac{n}{k}\, s(\lambda^S) \tag{2.27}$$

which, obviously, is a consequence of combining Eq. 2.25 and Eq. 2.26.

The previous formula can be illustrated from Table 2.1, for the 12-TET system, as follows. Each mode class with 3 shifts generates $n = 12$ trichords. These chords are obtained by applying one particular shift of the class to every note of the 12-TET system. Then, each mode class with 3 elements generates 12 trichords, and there are 18 mode classes of this type. On the other hand, the unique mode class with one shift is $[4,4,4]$, which generates $\frac{12}{3} = 4$ different trichords. These are, for instance, the four directed chords resulting from applying the mode $[4,4,4]$ to the first four consecutive notes of the 12-TET system. Therefore, in total, there are $12 \cdot 18 + 4 = 220$ trichords.

In a similar way, we may count the number of hexachords in the 12-TET system. We write the terms of Eq. 2.27 for $n = 12$, $k = 6$. According to the previous section there are 75 mode classes with 6 shifts, 3 mode classes with 3 shifts, 1 mode class with 2 shifts, and 1 mode class with 1 shift. Therefore, the number of hexachords is $75{\cdot}12+3{\cdot}6+1{\cdot}4+1{\cdot}2 = 924$. Notice that the second factor of each term is the number of different roots that generate different chords when a particular mode representing its mode class is used. This number is the size of the periodic pattern of the mode class, i.e., for $[1,1,4,1,1,4]$, $[1,2,3,1,2,3]$, and $[1,3,2,1,3,2]$ this size is 6, for $[1,3,1,3,1,3]$ is 4, and for $[2,2,2,2,2,2]$ is 2.

Theses enumerations of families of chords and modes have been done by applying basic combinatorics. There are specific and advanced mathematical tools, such as Burnside's and Redfield–Pólya's theorems, allowing to generalize such a procedure when particular families of objects can be defined according to their behavior with regard to some transformations. For example, a k-chord was defined as a subset of k notes, regardless their order. Therefore, a k-chord is any family of k-tuples of notes which is invariant or closed with regard to permutations. If a k-chord is defined from directed chords, i.e., a k-tuple of notes arranged clockwise direction on the octave, then a chord is a family invariant under rotations, either direct or inverse. Similarly, a mode class was defined as a family of modes which is invariant with regard to shifts. Then there is a group G that acts on a set C by transforming their elements, such as symmetry transformations, and particular families of elements in C

are invariant with regard to specific transformations of G. For chords, we shall see in the following sections that these transformations may act either on the root (translation, inversion, ...) or on the mode (shift, retrogradation, inversion, ...) and will have a group structure. Redfield–Pólya's theorem allows to enumerate in a general way the classes resulting from the action of the group G on C. In [21] these techniques were thoroughly used to enumerate several types of music objects.

CHAPTER 3

Subchords

3.1 UNIQUENESS OF CHORDS

A chord can be defined from several directed chords generated from any mode of a mode class applied to a particular root. If the directed chords are written clockwise direction along the circle of the octave, by starting at any note of the chord, the difference between two directed chords defining the same chord is only a rigid rotation of notes. This is given through the following condition.

▷ Two directed chords $a_0 \mid \mu$, $b_0 \mid \nu \in K(n, k)$ produce the same chord $\{a_0 \mid \mu\} = \{b_0 \mid \nu\}$ if and only if

$$\exists m \in \mathbb{Z}_k \text{ so that } b_i = a_{m+i}, \ \forall i \in \mathbb{Z}_k \tag{3.1}$$

which is equivalent to the following condition,

▷ Two directed chords $a_0 \mid \mu$, $b_0 \mid \nu \in K(n, k)$ produce the same chord $\{a_0 \mid \mu\} = \{b_0 \mid \nu\}$ if and only if

$$\exists m \in \mathbb{Z}_k \text{ so that } b_0 = a_m \text{ and } \nu = \mu^m \tag{3.2}$$

Notice that the general condition of ν and μ belonging to the same mode class is not sufficient, although $b_0 = a_m$ is satisfied. We see some examples:

1. The directed chords $a^0 = 1|[1, 3, 8] = (1, 2, 5)$ and $b^0 = 2|[3, 8, 1] = (2, 5, 1)$ of $K(12, 3)$ define the same chord $\{1, 2, 5\}$. The first rotation $a^1 = 2|[3, 8, 1]$ provides the value $m = 1$ such that $a_1 = b_0$ and $\nu = \mu^1$.

DOI: 10.1201/9781003597964-3

2. If we apply a translation by p notes to the root of both directed chords of the first example, we obtain the directed chords $a^0 = (p+1)|[1,3,8] = (p+1, p+2, p+5)$ and $b^0 = (p+2)|[3,8,1] = (p+2, p+5, p+1)$. Then, for the same value $m = 1$, as in the first example, $a_1 = b_0$ and $\nu = \mu^1$ are hold, so that both directed chords still define the same chord. This is true regardless the mode.

3. We consider a chord composed of the complementary pith classes of the previous chord, that is, $\{7, 10, 11\}$, and write it from roots that are the complementary of the ones of the previous example. Thus, $a^0 = 11|[8,3,1] = (11, 7, 10)$ and $b^0 = 10|[1,8,3] = (10, 11, 7)$. The condition of Eq. 3.2 is satisfied for the shift $i = 2$, being $a_2 = b_0$ and $\nu = \mu^2$.

4. We now give an example of an operation on the directed chords of the first example that does not produce equivalent chords. In each directed chord we change the root to its complementary PC by maintaining the mode. Thus, $a^0 = 11|[1,3,8] = (11, 0, 3)$ and $b^0 = 10|[3,8,1] = (10, 1, 9)$. None of the directed chords have a common note, hence they do not define the same chord.

3.2 INVERTED MODES

While any positive full mode $\mu = [A_0, A_1, \ldots, A_{k-1}] \in M(n,k)$ is clockwise directed, a *negative mode*

$$-\mu = [-A_0, -A_1, \ldots, -A_{k-1}]$$

also referred to as *inverted mode* of μ, defines a symmetric partition of the octave, though *anticlockwise direction*, when it is applied to the same root.

The negative mode $-[A_0, A_1, \ldots, A_{k-1}]$ defines the same partition of the octave that the composition of positive modes $[n-A_0][n-A_1]\cdots[n-A_{k-1}]$. But $[n-A_0, n-A_1, \ldots, n-A_{k-1}]$ is not a full mode, since $(n-A_0)+(n-A_1)+\cdots+(n-A_{k-1}) = (k-1)\,n$, instead of equal n. It is a generalized mode. Anticlockwise direction, however, the intervals of the negative mode $[-A_0, -A_1, \ldots, -A_{k-1}]$ do equal $-n$.

Therefore, we may speak of a *negative full* mode in the set $M(-n,k)$ when it satisfies such a condition. In this case, when it is applied to a root, a negative mode determines an anticlockwise directed chord. In this case, we may speak of opposite or *antidirected* k-chords, that have similar properties than the directed chords with regard to the chords they generate[1].

On the other hand, the full mode

$$\overline{\mu} = [A_{k-1}, \dots, A_1, A_0,]$$

describes a chord with similar intervals than $-\mu = [-A_0, -A_1, \dots, -A_{k-1}]$, when it is applied to the same root, although in clockwise direction. Therefore,

▷ The positive inverted full mode $\overline{\mu} = [A_{k-1}, \dots, A_1, A_0]$ is the reduced form of the negative inverted full mode $-\mu = [-A_0, -A_1, \dots, -A_{k-1}]$, i.e.,

$$\overline{\mu} = R(-\mu)$$

Let us prove it. To determine the relative partition induced in the octave by the mode $-\mu$ (except rigid rotations) we apply the generalized mode to the root 0. Bearing in mind that the addition of the positive mode intervals equals n and that any integer number of full octaves can be ignored, we write the notes determining the partition, which, in the end, define a directed chord with the notes arranged clockwise,

$$\begin{aligned}
&0|[n - A_0, n - A_1, \dots, n - A_{k-1}] = \\
&(0, n - A_0, 2n - A_0 - A_1, \dots, kn - A_0 - A_1 - \cdots - A_{k-2}) = \\
&(0, n - A_0, n - A_0 - A_1, \dots, n - A_0 - A_1 - \cdots - A_{k-2}) = \\
&(0, \cancel{A_0} + A_1 + \cdots + A_{k-1}, \cancel{A_0} + \cancel{A_1} + A_2, \cdots + A_{k-1}, \dots, \cancel{A_0} + \cancel{A_1} + \cdots + \cancel{A_{k-2}} \\
&+A_{k-1}) = (0, A_{k-1}, A_{k-1} + A_{k-2}, \dots, A_{k-1} + A_{k-2} + \cdots + A_0) = 0| \\
&[A_{k-1}, \dots, A_1, A_0]
\end{aligned}$$

Hence, the latter full mode is equivalent to the former generalized mode. Therefore,

▷ The directed mode μ generating a directed chord $a_0 \,|\, \mu$, is the reduced form of its *retrograde* mode $-\overline{\mu}$, so that, applied anticlockwise direction to a_0, generates the same chord,

$$\mu = R(-\overline{\mu}) \; ; \quad \{a_0 \,|\, -\overline{\mu}\} = \{a_0 \,|\, \mu\} \tag{3.3}$$

The mode class of a negative full mode is the one corresponding to its positive form.

[1]They can be associated with an arpeggio that remains within the octave beginning with the root, with their notes decreasing in pitch.

3.3 RELATIVE INVERTED CHORDS

The retrograde mode of $\overline{\mu}$ is $-\mu$, so that the inverted mode of μ has two expressions: the positive mode $\overline{\mu}$ and the negative mode $-\mu$. If they are applied to the root a_0, we get two directed chords: the *relative inverted*, $a_0 \,|\, \overline{\mu}$, read clockwise; and the *relative mirror* $a_0 \,|\, -\mu$, read anticlockwise. Both are about a_0 and define the same chord

$$\{a_0 \,|\, \overline{\mu}\} = \{a_0 \,|\, -\mu\} \tag{3.4}$$

These expressions determine the *relative inverted chord* of the chord $\{a_0 \,|\, \mu\}$. From a geometrical viewpoint, it is the symmetric chord of $\{a_0 \,|\, \mu\}$ with regard to the root, i.e., a reflection.

Nevertheless, if we write the directed chord $a_0 \,|\, \mu$ from another root to generate the same chord $\{a_0 \,|\, \mu\}$, such as the rotation $a_i \,|\, \mu^i$, its relative mirror chord $\{a_i \,|\, -\mu^i\}$ is different from $\{a_0 \,|\, -\mu\}$. In other words, the relative mirror chord depends on the directed chord we are using to express the chord. This also applies to the relative inverted chord. Therefore, the relative inversion is not an operation well defined on the class of chords, it is only for directed chords.

In general, two directed chords $x \,|\, \mu$ and $y \,|\, -\mu$ are mutual *reflection* about $\frac{x+y}{2}$, because added by components their semi-sum accounts $\frac{x+y}{2}$ in each component. However, it is possible that $\frac{x+y}{2}$ does not belong to $\mathbb{Z}_n$, i.e., is not a note of the n-TET system. The way to seek whether two directed chords $(a_0+x) \,|\, \mu$ and $(a_0-x) \,|\, \nu$, generated by positive modes, are mutual reflection about a_0 is to express one of them as generated by a negative mode, $(a_0 - x) \,|\, \nu = (a_0 - x) \,|\, -\overline{\nu}$, i.e., if $\overline{\nu} = \mu$, then they are. In general, the way to seek whether two directed chords $a_0 \,|\, \mu$ and $b_0 \,|\, \nu$, generated by positive modes, are mutual reflection is to rotate one directed chord and to find a shift ν^i so that $\nu^i = \overline{\mu}$. Then, these directed chords are symmetric about $\frac{a_0+b_i}{2}$.

3.4 INVERTED CHORDS

If $-\mu$ is applied to the complementary PC of a_0, $-a_0 \equiv -a_0 \mod n$, all the notes of the resulting chord $\{-a_0 \,|\, -\mu\}$ are complementary PCs of the ones in $\{a_0 \,|\, \mu\}$.

Then, $\{-a_0 \,|\, -\mu\} = \{-a_0 \,|\, \overline{\mu}\}$ is referred to as the *inverted chord* of $\{a_0 \,|\, \mu\}$.

Obviously, such an inversion of root and mode is defined for directed chords. The *mirror-directed chord* of $a_0 \,|\, \mu$ is expressed as $-a_0 \,|\, -\mu$, from the negative mode, by writing

$$-(a_0 \,|\, \mu) = -a_0 \,|\, {-\mu} \tag{3.5}$$

and the *inverted directed chord* is written as $-a_0 \,|\, \overline{\mu}$, from the positive inversion of the mode. According to Eq. 3.4, both expressions of the inverted directed chord generate the same chord.

▷ The negative inversion of two directed chords, namely $a_0 \,|\, \mu$ and $b_0 \,|\, \nu$, that define the same chord

$$\{a_0 \,|\, \mu\} = \{b_0 \,|\, \nu\}$$

produce two similar chords, i.e.,

$$\{-a_0 \,|\, {-\mu}\} = \{-b_0 \,|\, {-\nu}\}$$

It can be easily deduced from the condition given in Eq. 3.1. We know that it exists a value $m \in \mathbb{Z}_k$ such that $(a_m, \ldots, a_{m-1}) = (b_0, \ldots, b_{k-1})$, i.e., both directed chords match. Then, by taking complementary mode classes on both sides and arranging them clockwise along the octave, we get $(-a_{m-1}, \ldots, -a_m) = (-b_{k-1}, \ldots, -b_0)$. Hence both mirror-directed chords define the same inverted chord.

We may then speak of inversion of a chord as the result of applying the inversion or mirroring to any directed chord producing the same chord, that is,

$$-\{a_0 \,|\, \mu\} = \{-a_0 \,|\, {-\mu}\} = \{-a_0 \,|\, \overline{\mu}\} \tag{3.6}$$

so that, the inversion is an operation well defined for chords as well.

In Table 3.1, some definitions of directed chords obtained from the modes $-\mu$ and $\overline{\mu}$ are listed together with their symmetry properties. They are displayed by couples of directed chords that generate the same chord.

3.5 STRUCTURE OF CHORDS

By expressing a directed chord $a = a_0 \,|\, \mu$ as a mode applied to a root, we are able to describe easily several well-known families of chords. The simplest example is the family of chords obtained by maintaining the mode μ and by translating the root a_0 to any note of the scale. These

Table 3.1 Some related directed chords. The directed chords among two consecutive horizontal lines produce the same chord.

Directed chords related to $a \mid \mu$		Symmetry point
Mirror	$-a \mid -\mu$	0
Inverted	$-a \mid \overline{\mu}$	0
Mirror by x	$(x-a) \mid -\mu$	$x/2$
Inverted by x	$(x-a) \mid \overline{\mu}$	$x/2$
Relative mirror	$a \mid -\mu$	a
Relative inverted	$a \mid \overline{\mu}$	a
Relative mirror by x	$(x+a) \mid -\mu$	$a + x/2$
Relative inverted by x	$(x+a) \mid \overline{\mu}$	$a + x/2$

directed chords have the same interval structure, although they differ in PCs, therefore in timbre.

Another example is the family of rotations of a directed chord $a = a_0 \mid \mu$. Indeed, we had defined a chord as the class of directed chords with regard to chord rotations. This family is composed of the directed chords produced by the shifts μ^i of the mode μ applied to the notes a_i of the initial directed chord. If the directed chords are considered in the octave that begins at each root, they sound very similar.

A third example is the family of directed chords obtained by maintaining the root a_0 and applying different shifts μ^i of the mode. Since the root is the lower pitch of a directed chord, the ear will give us a specific perception of the other tones in contrast to the partials of the root. For example, the three shifts of the mode $\mu = [4, 3, 5]$ applied to the root C produce the C major chord, the first rotation of the A$^\flat$ major chord, and the second rotation of the F major chord, all of them beginning with the note C.

In general, for a mode $\mu = [A_0, \ldots, A_{k-1}] \in M(n,k)$, we notate $\mathcal{S}(\mu)$ the family of its *symmetric modes*, that is, the set of modes containing all the permutations of the intervals of μ. Hence, $\mathcal{S}(\mu)$ is a class of modes within $\mathcal{M}(n,k)$ that contains the classes with regard to shifts, $\mathcal{C}(\mu)$, and, in particular, it contains the shifts of μ, i.e., the class μ^S, and the shifts of its inverted mode, i.e., the class $\overline{\mu}^S$. The chords generated from symmetric modes applied to the same root have particular musical interest.

For instance, the basic harmonic structure of trichords in tonal music is based in the following facts. In the 12-TET scale, for the three positive intervals $A, B, C \in \mathbb{Z}$ that generate a trichord, satisfying $A+B+C = 12$, we ask to fulfill: (1) the whole 12-TET scale should be obtained by applying recursively at least one of them, and (2) each interval should be greater than one tone, i.e., $A, B, C > 2$.

Since the 12-TET scale has the structure of the cyclic group $(\mathbb{Z}_{12}, +)$, their generators are 1,5,7,11, coprime with 12. Then, from the second requirement we must exclude 1, 7, and 11. Therefore, one of the intervals is 5. Hence, its complementary interval must be decomposed as $3 + 4$. Thus, the only possible three intervals of the mode are 3, 4, and 5. Then, the symmetric modes of $\mu = [3, 4, 5]$ (that for $k = 3$ include only two mode classes μ^S and $\overline{\mu}^S$, i.e., those of $[3, 4, 5]$ and $[4, 3, 5]$, with their shifts) applied to one root generate all the major and minor trichords.

In addition, the notes provided by the mode $[4, 3, 5]$ applied to one root are the lower ones and the more significant of the harmonic series, i.e., the fundamental, the third harmonic and the fifth harmonic, which form the major triad. Somehow, it can be said that this fact is the starting point of two ways of analyzing music, one, Schenkerian theory for the deep structure of tonal music, and the other, Neo-Riemannian theory for transformations and voice leading over the Tonnetz, which can be extended to non-tonal music.

Let us remember that the 12-TET scale is very close to the 12-tone Pythagorean scale, with the same structure of cyclic group $(\mathbb{Z}_{12}, +)$. Its generators are the iterations of fifths that are coprime with 12, i.e., $\nu_1, \nu_5, \nu_7, \nu_{11}$, with respective ordinals $\vartheta_7, \vartheta_{11}, \vartheta_1, \vartheta_5$. Therefore, according to Section 1.5, the interval 5 of the 12-TET scale corresponds to ν_{11} (the fourth) and the interval 7 to ν_1 (the fifth). With regard to the subdivisions of the interval 7, the interval 3 corresponds to ν_9 (the minor third), and 4 to ν_4 (the major third).

As seen in Section 1.6, the structure induced by the mode intervals is correlated to the harmonic consonance of chords and defines which are neighbor chords with regard to the root. The change toward other tonalities is carried out by translations according to this topography. The cases explained above will be studied carefully in the following chapters, either as transformations of the root, transformations of the mode, or both. From a mathematical point of view, these transformations are group actions on the root and mode.

3.6 INVARIANT SUBCHORDS

In the following sections we shall refer to a chord $\{a\} \in \mathcal{K}(n,p)$ as the directed chord $a \in K(n,p)$ with its notes ordered from 0 to $n-1$ along the circle of the octave. The chord notation $\{a\}$ will be just used to point out that the elements of the directed chord a are considered regardless the order.

In a n-TET system, starting from a directed p-chord $a = (a_0, \ldots, a_{p-1}) \in K(n,p)$, we shall generate a new directed chord having a fixed set of $(q+1)$ consecutive notes of a, $q+1 < p$.

First, we fix the first $(q+1)$ notes of a, and write a as the union of two directed subchords $b \in K(n, q+1)$ and $c \in K(n, p-q-1)$, according to

$$\{a\} \equiv \{b\} \cup \{c\} = \{a_0, \ldots, a_q\} \cup \{a_{q+1}, \ldots, a_{p-1}\}$$

where the first directed subchord b contains the fixed notes.

We now express the directed chord a as generated by a full mode $\alpha = [A_0, \ldots, A_{p-1}] \in M(n,p)$,

$$a = a_0 \,|\, \alpha$$

and write the mode α as the composition

$$\alpha = \beta \cdot \gamma$$

of the partial modes $\beta = [A_0, \ldots, A_{q-1}] \in M(m,q)$ and $\gamma = [A_q, \ldots, A_{p-1}] \in M(n-m, p-q)$, which are complementary in $M(n,p)$. Then, we have

$$a = a_0 \,|\, \beta \cdot \gamma \tag{3.7}$$

According to Eq. 2.20, the chord is obtained as

$$\{a\} = \{a_0 \,|\, [\beta, *]\} \cup \{a_0 \,|\, [*, \gamma]\}$$

Thus, the partial mode β applied to the note a_0 secures the invariant notes of the first subchord. Therefore, for a mode $\beta \in M(m,q)$, the family of modes in $M(n,p)$

$$O_\beta = \{\beta \cdot \kappa;\ \kappa \in M(n-m, p-q)\} \tag{3.8}$$

leaves invariant the directed $(q+1)$-chord

$$b = a_0 \,|\, \beta$$

as a part of the directed p-chord a.

The family of modes O_β is referred to as the *orbit* of β. Any supermode of the 1-mode $[n-m]$ is complementary of β in $M(n,p)$ and the mode $\beta \cdot [n-m]$ is the less fine mode of O_β.

3.7 SYMMETRIC MODES

For a mode $\alpha \in M(n,p)$, the set of the symmetric modes of α, i.e., those obtained from any permutation of the intervals composing α, is denoted as $\mathcal{S}(\alpha)$. The cardinality of $\mathcal{S}(\alpha)$ depends on the number of equal intervals in α. If they are all different, $\#\mathcal{S}(\alpha) = p!$. If α has $q \leq p$ intervals A_i; $i = 0, \ldots, q-1$, each one repeated n_i times, so that $\sum_{i=0}^{q-1} n_i = p$, then

$$\#\mathcal{S}(\alpha) = \frac{p!}{n_0! \cdots n_{q-1}!} \tag{3.9}$$

On the other hand, it is possible to compute the cardinality of $\mathcal{S}(\alpha)$ from the mode classes $\mathcal{C}(\alpha)$ it contains. Then, since $\mathcal{S}(\alpha) = \alpha_1^S \cup \cdots \cup \alpha_m^S$, $\alpha_i^S \in \mathcal{C}(\alpha)$, by taking into account the number of shifts in each mode class, we have $\#\mathcal{S}(\alpha) = \sum_{i=1}^m s(\alpha_i^S)$.

The chords generated by the modes of $\mathcal{S}(\alpha)$, in particular when they are applied to the same root, have a sound quality related to the chord generated by α. Thus, for the mode $\beta \in M(m,q)$, with $m < n, q < p$, we shall take into account the subset of O_β formed by the modes $\lambda = \beta \cdot \gamma'$, with $\gamma' \in \mathcal{S}(\gamma)$. We shall refer to those modes within the orbit of β as symmetric modes with regard to the partial mode γ, which form the *symmetric orbit* of O_β, denoted as $O_{\beta,\gamma} \subset \mathcal{S}(\alpha)$.

Then, the number of chords in the form $a = a_0 \,|\, \beta \cdot \gamma'$ that share the first $q+1$ notes is $(p-q)!$, since $\gamma' \in K(n-m, p-q)$.

Furthermore, by taking into account the shifts of the modes containing the partial mode β, we may define the extended orbit of β as

$$O'_\beta = \{(\beta \cdot \kappa)^i;\ \kappa \in M(n-m, p-q);\ i = 0, \ldots, n-1\} \tag{3.10}$$

Similarly, we may consider the extended symmetric orbit $O'_{\beta,\gamma}$ that contains the shifts of the modes in $O_{\beta,\gamma}$.

As explained in the previous section, for $\beta \in M(n,p)$ and by writing $x_0 = a_0$, the directed chords

$$x = x_0 \,|\, \beta \cdot \kappa; \quad \kappa \in M(n-m, p-q) \tag{3.11}$$

generated by the modes of O_β, leave invariant the first subchord $\{x_0 \,|\, \beta\} = \{x_0, \ldots, x_q\}$ of the original directed chord $x_0 \,|\, \beta \cdot \gamma$, $\gamma \in M(n,p)$. Then, any rotation of x, such as

$$x^i = x_i \,|\, (\beta \cdot \kappa)^i \tag{3.12}$$

gives the same chord $\{x\}$, and also fixes the same notes $\{x_0, \ldots, x_q\}$.

Therefore, the modes of the extended orbit O'_β leave the subchord $\{x_0 \,|\, \beta\}$ invariant when applied to the appropriate root x_i.

On the other hand, in order to fix the $q+1$ notes of the directed chord $r = r_0 \,|\, \rho \in K(n,p)$ by beginning at the note r_i, $0 < i \leq p-1$, we express the directed chord r from its i-th rotation, such as

$$r^i = r_i \,|\, \rho^i \tag{3.13}$$

Then, by identifying root and mode in $x_0 \,|\, \beta \cdot \gamma \equiv r_i \,|\, \rho^i$, we apply the decomposition $\rho^i = \beta \cdot \kappa$ as explained above to get the chords with the invariant subchord.

Finally, as a particular case of Eq. 3.11, the chords generated from the symmetric modes of the initial chord of Eq. 3.7 that leave $x_0 \,|\, \beta$ invariant are obtained as $x_0 \,|\, \beta \cdot \gamma'$ and its rotations, which are generated in a similar way by the modes of the symmetric extended orbit $O'_{\beta,\gamma}$.

3.8 TRICHORD EXAMPLES

We show some basic examples of the previous concepts for trimodes and the trichords they generate. We shall notate with the symbol $\sim$ when a directed chord a or its rotations a^i belong to the class of the chord $\{a\}$,

$$\{a\} \sim a \;, \quad \{a\} \sim a^i$$

It also means that in the class we take the specific directed chord as representative to work with it.

The trichord $r = \{0, 10, 30\} \in \mathcal{K}(53, 3)$ is obtained from three equivalent modes as

$$r = \{0, 10, 30\} \sim 0 \,|\, [10, 20, 23] \sim 10 \,|\, [20, 23, 10] \sim 30 \,|\, [23, 10, 20]$$

Suppose we wish to obtain another chord that maintains all the notes but one, e.g., the first two notes 0 and 10. We write $r \sim 0 \,|\, [10][20, 23]$, so that, according to Eq. 3.8, all the modes from the orbit $O_{[10]}$ applied to the note 0 satisfy such a condition. Thus, any chord $s \sim 0 \,|\, [10][M, N]$, with $M + N = 43$, maintains the subchord $\{0, 10\}$ of the initial chord.

Moreover, to compare to other chords generated by the same intervals, we center our attention in the symmetric modes of $[10, 20, 23]$ within the orbit of $[10]$, that is, the symmetric orbit $O_{[10],[20,23]} = \{[10][20, 23], [10][23, 20]\}$. The new mode $[10, 23, 20]$ generates the chord

$$s = \{0, 10, 33\} \sim 0 \,|\, [10, 23, 20] \sim 10 \,|\, [23, 20, 10] \sim 33 \,|\, [20, 10, 23]$$

along with its rotations.

Alternatively, we may want to fix the subchord $\{10, 30\}$ of the chord $r = \{0, 10, 30\}$. According to Eq. 3.13, we write it from the first rotation $r^1 = \{10, 30, 0\} \sim 10 \,|\, [20, 23, 10] \sim 10 \,|\, [20][23, 10]$. Then, by shifting the partial mode $[23, 10]$, we get the new chord, with the same intervals,

$$s' = \{10, 30, 40\} \sim 10 \,|\, [20, 10, 23] \sim 30 \,|\, [10, 23, 20] \sim 40 \,|\, [23, 20, 10]$$

expressed according to the three rotations.

Similarly, to fix the notes $\{30, 0\}$ of the chord r, we work out the second rotation $r^2 = \{30, 0, 10\} \sim 30 \,|\, [23, 10, 20]$ and, by transposing the two last intervals, obtain

$$s'' = \{30, 0, 20\} \sim 30 \,|\, [23, 20, 10] \sim 0 \,|\, [20, 10, 23] \sim 20 \,|\, [10, 23, 20]$$

according to the three rotations.

3.9 TRICHORDS SHARING TWO NOTES

In general, for a trichord

$$\{a, b, c\} \sim a \,|\, [A, B, C] \sim b \,|\, [B, C, A] \sim c \,|\, [C, A, B] \tag{3.14}$$

there are at most three chords (with the corresponding rotations) generated by the symmetric modes of any rotation of the initial chord that maintain two of the notes of the initial chord:

$$\begin{aligned} &\{a, b, d\} \sim a \,|\, [A, C, B] \sim b \,|\, [C, B, A] \sim d \,|\, [B, A, C] \\ &\{b, c, e\} \sim b \,|\, [B, A, C] \sim c \,|\, [A, C, B] \sim e \,|\, [C, B, A] \\ &\{c, a, f\} \sim c \,|\, [C, B, A] \sim a \,|\, [B, A, C] \sim f \,|\, [A, C, B] \end{aligned} \tag{3.15}$$

All the symmetric modes of $[A, B, C]$ appearing in Eq. 3.15 belong to the mode class of $[A, C, B]$, obtained by applying one interval transposition to the initial mode, while those of Eq. 3.14 belong to the mode class of $[A, B, C]$.

If the intervals A, B, C are non-two equal, the above three chords are different. If there is a couple of similar intervals, in Eq. 3.15 there are two different chords. If the three intervals are equal, then the initial chord $\{a, b, c\}$ does not change.

The chords and modes of Eqs. 3.14 and 3.15 are shown in Figure 3.1, where the symmetric modes of $[A, B, C]$, applied to the appropriate root, leave invariant each one of the 2-subchords.

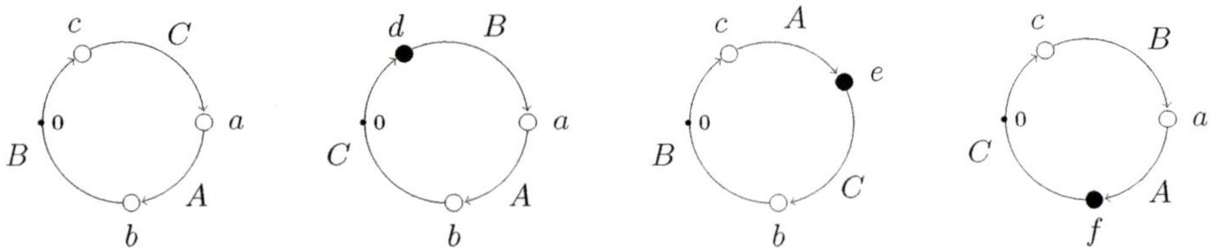

Figure 3.1 Trichord $\{a, b, c\} \sim a \,|\, [A, B, C]$ and chords obtained from the symmetric modes containing two of the notes of the initial chord.

In addition, from the Eqs. 3.14 and 3.15, we see that it is also possible to express any chord as generated by any one of the shifts of the mode class.

The symmetric trimodes may be described as the action of the symmetric group $\mathcal{S}_3$ on the set of the intervals $\{A, B, C\}$.

In $O_{[A]}$, the symmetric modes in regard to $[B, C]$ are $O_{[A],[B,C]} = \{[A, B, C], [A, C, B]\}$; in $O_{[B]}$, the symmetric modes in regard to $[A, C]$ are $O_{[B],[A,C]} = \{[B, A, C], [B, C, A]\}$; and in $O_{[C]}$, the symmetric modes in regard to $[A, B]$ are $O_{[C],[A,B]} = \{[C, A, B], [C, B, A]\}$.

In Figure 3.2 the relationship between these modes is displayed. The vertices connected with red arrows belong to the same mode class. The red arrow represents a single shift, given by the permutation $(3, 1, 2)$. The vertices of the triangular diagrams connected with a blue line are the subsets $O_{[A],[B,C]}$, $O_{[B],[A,C]}$, $O_{[C],[A,B]}$. The blue line represents the permutation $(1, 3, 2)$, i.e., a transposition of two intervals. The modes of the external triangle, applied as in Eq. 3.14, generate the chord $\{a, b, c\}$. The modes of the internal triangle, applied as in Eq. 3.15, generate the

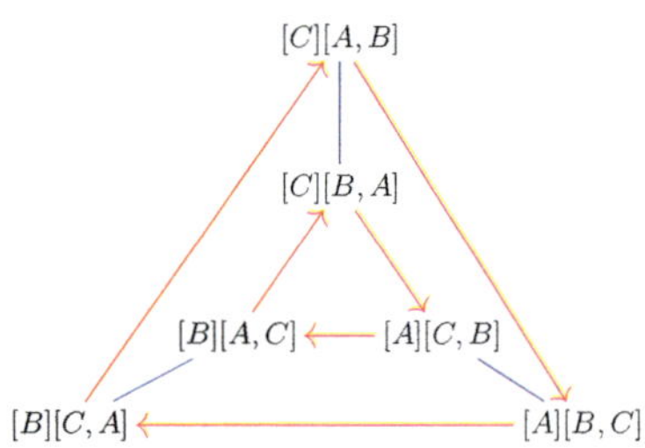

Figure 3.2 Cayley graph for the symmetric modes with single intervals A, B, C, non-two equals.

chords sharing two notes with the initial chord. Such a diagram, when generalized to higher-order modes, is equivalent to the Cayley graph of the respective symmetric group. Similarly, it can be generalized to n-chords to determine the neighbor chords that share $n - 1$ notes.

CHAPTER 4

Trichords

4.1 CHORD EXTENSION [A,B,C]

The properties of trichords will be now studied from a simple, geometric approach. In the next chapters these properties will be generalized and derived from an algebraic approach.

The geometric relationship among the initial chord and the three different chords sharing two notes with it is displayed in the graph of Figure 4.1. This is the *extension* of the chord, containing all the symmetric modes.

The central triangle of the extension contains the initial chord $\{a, b, c\}$, generated by the mode $[A, B, C]$, that we first assume with non-two equal intervals. It is surrounded by three triangles sharing an edge with it, corresponding to the chords generated by the shifts of the class $[A, C, B]$.

This structure may be repeated around each new triangle to form a tonal network, or *Tonnetz*, first described by Euler in 1739, originated in the chord $\{a, b, c\} \sim a\,|\,[A, B, C]$ with alternance of the two mode classes of the modes $[A, B, C]$ and $[A, C, B]$. The modes of the first class are clockwise directed in the triangles shaped as ∇, while the modes of the second class are anticlockwise directed in the triangles shaped as Δ. For trimodes, we shall say that the former are positive and the latter negative, although this cannot be generalized to higher-order modes. For instance, the symmetric modes of a tetrachord have up to 6 mode classes.

Since $A + B + C = n$, there are actually two degrees of freedom in constructing such a network, allowing us to draw those triangles on a surface. One simple representation consists in, starting with the note a, to take increments of $A \bmod n$ units along the horizontal direction rightward, and of $B \bmod n$ units vertically downward, obtaining thus an

DOI: 10.1201/9781003597964-4

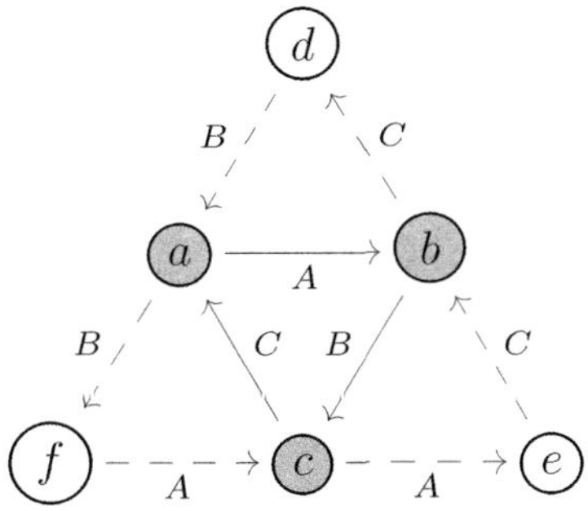

Figure 4.1 Extension $[A, B, C]$ formed by the chords sharing two notes with the initial chord $a \mid [A, B, C]$ (notes in gray). The three new chords around it are generated by the symmetric modes of a different mode class.

unbounded bidimensional array of nodes with coordinates (iA, jB) mod n; $i, j \in \mathbb{Z}$, with the center $(0, 0)$ in a, that are associated with the notes

$$m_{ij} = a + iA + jB \mod n; \quad i, j \in \mathbb{Z} \tag{4.1}$$

By this way, when closing one triangle, which is equivalent to adding an interval of C units, the circle of the octave becomes also closed.

The tone network is periodic in both *main directions*, and diagonally as well, although the full set of notes of the n-TET system arise in any direction only when the corresponding interval is coprime with n. This structure is equivalent to a Cartesian product of two circles, one for the notes iA mod n and the other one for the notes jB mod n, so that the tonal network can be drawn on a torus. In Section 5.2, this structure will be studied more rigorously.

More precisely, the whole notes $k = 0, \ldots, n-1$ of the tonal system are represented in the torus if, and only if, $k = a + xA + yB + zn; \quad x, y, z \in \mathbb{Z}$ is fulfilled. We may clearly assume the origin at $a = 0$. Since the three intervals equal n, the above relationship can be written as

$$k = pA + qB + rC; \quad p, q, r \in \mathbb{Z}\,; \; k = 0, \ldots, n-1 \tag{4.2}$$

Therefore, according to Bézout's identity for three integers, in order to express all the notes from any of the intervals, it is required that

$\gcd(A, B, C) = 1$. In general, if $\gcd(A, B, C) = d$, then d is the smallest positive integer that can be expressed in the combination of Eq. 4.2, which provides a number of $v = \frac{n}{d}$ different notes and vertices for the Tonnetz.

Since each mode class generates a unique chord for every note of the n-TET system, if the Tonnetz has v vertices, it is then composed of $f = 2v$ different trichords, which are the faces of the regular tessellation of the torus with triangles. It is worth noticing that three vertices define a face of the triangle only when they are connected by three edges corresponding to the three consecutive mode intervals (ranging one octave), all of them with either positive orientation or negative orientation. Positive and negative alternated intervals do not form a loop.

The number of edges in the tessellation is $e = \frac{3}{2}f = 3v$, since each triangle has three edges and each edge is shared by two triangles. Thus, we obtain the Euler characteristic of a torus,

$$\chi = v - e + f = 0$$

4.2 EXTENSION WITH A NEW CHORD

There is a particular case where the vertices of the extension, i.e., the notes $\{d, e, f\}$ in Eq. 3.15, are non-two equal and are different from the notes in $\{a, b, c\}$. Then, they will form a new trichord. These notes are new ones only in the following cases

$$\begin{aligned} d \neq c &\iff B \neq C \\ e \neq a &\iff A \neq C \\ f \neq b &\iff A \neq B \end{aligned} \tag{4.3}$$

Assuming that these notes do not belong to the chord $\{a, b, c\}$, we check when they are non-two equal. From Figure 3.1, it is easy so see that

$$d = b + C = a + A + C \bmod n; \quad e = c + A = b + A + B \bmod n;$$
$$f = a + B = c + B + C \bmod n$$

Hence, their differences account

$$e - d = A + B - C \bmod n; \quad d - f = A - B + C \bmod n; \quad f - e = -A + B + C \bmod n$$

Since $A + B + C = n$, we obtain

$$
\begin{aligned}
e - f &= 2A \bmod n \\
f - d &= 2B \bmod n \qquad (4.4) \\
d - e &= 2C \bmod n
\end{aligned}
$$

Therefore, $\{d, e, f\}$ is a new trichord only in the case of non-two equal values of A, B, C, provided that those values are different from $\frac{n}{2}$, as is deduced from Eq. 4.4. We may then express the new trichord as

$$\{f, e, d\} \sim f \mid [2A, 2C, 2B] \bmod n \qquad (4.5)$$

Depending on the values A, B, C, and n, Eq. 4.5 allows us to write iteratively a sequence of chords of variable length starting from Eq. 3.14, at most of periodicity n, so that each chord has different notes than the previous one until ending the sequence. This algorithm and other similar strategies may be used to generate sequences of chords with no matching notes.

4.3 TONNETZ EXAMPLE

In order to draw the tonal network associated with a trimode $[A, B, C]$ with non-two equal intervals, we choose two coordinate axes in arbitrary directions, one in steps of A units, marked in blue in the example of Figure 4.1, and the other in steps of B units, marked in red, that determine the coordinates of the nodes of the network representing the notes. These nodes are then the vertices of a mosaic of parallelograms. Starting from one node, to advance one step in each positive direction of the axes is equivalent to advancing one step of $-C \bmod n$ units in the positive direction of the diagonal of the parallelogram, or C units in the negative direction of the diagonal, marked in green in the figure.

We assume that the center of the coordinate system is the note a. Then, according to Eq. 4.2, a node of coordinates (p, q) represents the note

$$k(p, q) = a + pA + qB \mod n$$

As the net is periodic in all directions, the note a will appear every time that

$$pA + qB \mod n = 0$$

This determines the coordinates of the replicas of the center of the net.

For the 53-TET scale, the tonal network associated with the trichord $\{0, 10, 30\} \sim 0\,|\,[10, 20, 23]$ is displayed in Tab. 4.1 . The intervals $A = 10$, $B = 20$, and $C = 10$ are represented by blue, red, and green arrows, respectively; not exactly directed as explained in the above section, for the sake of obtaining a more compact diagram. Moreover, a slight shear of the parallelogram makes easy to identify top and bottom sides, and right and left sides, to build up a torus. The initial chord is given in boldface. Repeated notes are displayed in gray, allowing to draw exactly the whole number of 106 triangles for the different chords. The non-repeated 53 vertices, 159 edges, and 106 triangles account for the Euler characteristic of a torus.

Notice that in any parallelogram formed by four connected neighbor vertices, a chord is not formed by choosing any three out of the four vertices. Since the orientation of the edges must be preserved, there are only two chords inside each parallelogram.

4.4 CHORD EXTENSION [A,B,A]

We assume a the center of the tonal network and, according to Eq. 4.2, write the notes of the network in terms of the axes corresponding to the intervals A (horizontal axis, in blue in Figure 4.2) and $-C$ (diagonal axis, in green), as

$$k(p, q) = a + pA + q(-C) \mod n$$

The network is periodic between points satisfying

$$pA - qC \mod n = 0$$

Now we assume that two of the intervals are similar, e.g., $C = A$, so that the initial chord is $a\,|\,[A, B, A]$. Thus, a node takes periodically the value a when

$$(p - q)A \mod n = 0$$

In particular, all the notes of the nodes satisfying $p = q$ match the note a.

The chord extension $[A, B, A]$ is displayed in Figure 4.2, where only two new chords and two new notes are obtained. Since all the symmetric modes of $[A, B, A]$ belong to a single mode class, if the number of different vertices in the tonal net is v, although the number of chords is still $f = 2v$, the number of different chords and faces is v. Thus, in this case the tonal network is equivalent to a tours with the half central part

Table 4.1 Tonnetz generated by the chord $\{0, 10, 30\} \sim 0 \,|\, [10, 20, 23]$ in a 53-TET system.

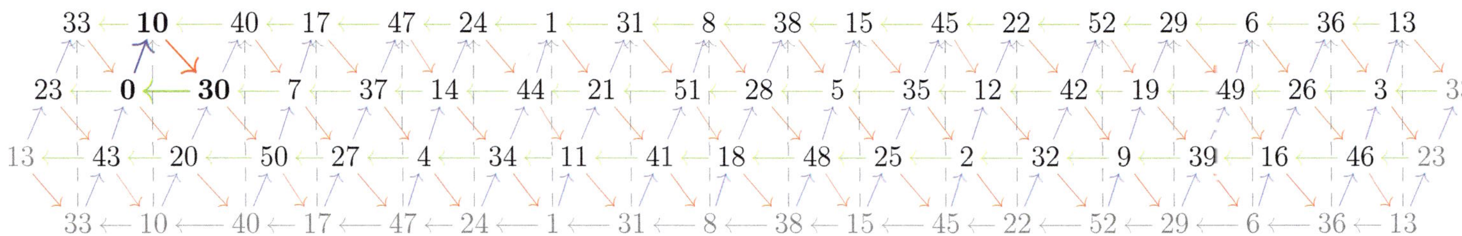

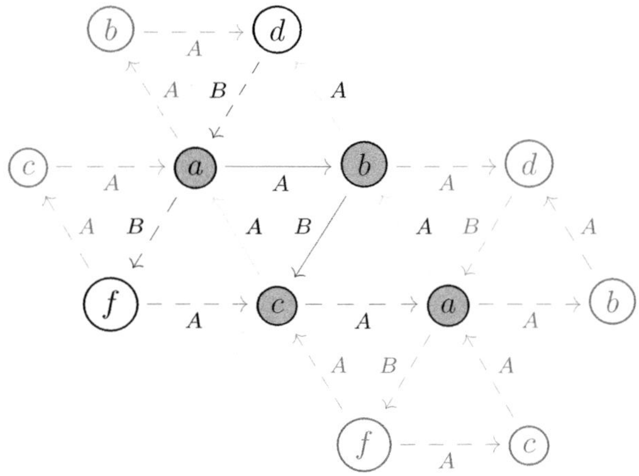

Figure 4.2 The chords sharing two notes with the initial chord $a \mid [A, B, A]$ (notes in gray) form the chord extension $[A, B, A]$, with two new chords generated by a single mode class. The prolongation of the chord extension (in gray) shows the symmetry and periodic properties of the tonal net.

having the same chord distribution as the half outer part. This will be called in Section 5.4 a degenerate Tonnetz, where an equivalence relation identifying similar notes converts the torus into a Möbius band.

4.5 CHORD EXTENSION [A,2A,A]

A particularly simple case of the previous degenerate chord extension occurs when one of the intervals satisfies one of the relationships in Eq. 4.4. Since we are assuming the case $A = C$, it is only possible $2B = n$. Thus, we have $A = C = \frac{n}{4}, B = 2A = \frac{n}{2}$, and $n = \dot{4}$. It is the chord extension $[A, 2A, A]$. Then, the nodes of the previous extension satisfy $d = f$ and, in the diagonals parallel to B, the notes separated $2B$ units are the same. Therefore, the chord extension $[A, B, A]$ of Figure 4.2 is reduced to Figure 4.3, with one new chord.

Then, the tonal network is only composed of four different chords and notes.

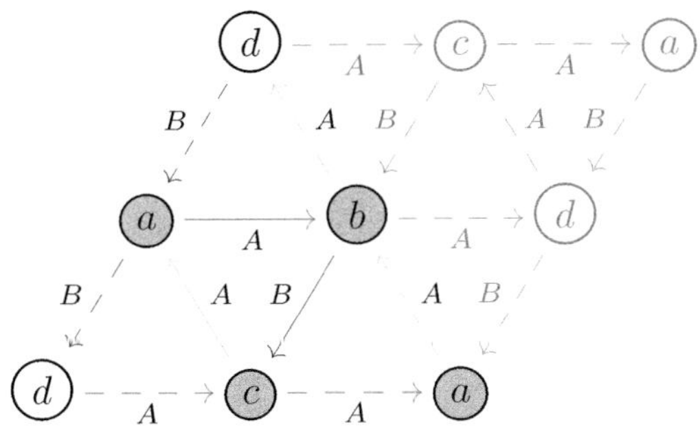

Figure 4.3 Extension of the chord $a \mid [A, B, A]$, with $B = 2A$. The prolongation of the chord extension (in gray) shows the full tonal network.

4.6 TONAL CELL [A,B,C]

From another viewpoint, the Tonnetz builds up a mosaic of hexagonal cells composed of six chords in the shape of triangles with a common vertex. These, chords are generated by the symmetric modes of the initial chord $a \mid [A, B, C]$. Each node of the network is the center of a cell, which contains the root, which is common note of the six chords of the *tonal cell*. In addition, these chords share another note (and the corresponding interval) with their neighbors. Therefore, the symmetric modes of the original mode $[A, B, C]$ determine a family of neighbor chords that, in musical terms, are harmonically related.

For the general case of an extension $[A, B, C]$, the hexagonal tonal cell shown on the left side of Figure 4.4 is composed of six different chords and six different symmetric modes. The clockwise directed modes are the modes of the external vertices of the graph in Figure 3.2, while anticlockwise's are internal's. On the other hand, we may use a simpler notation by assuming that the positive orientation of every mode in a triangle is clockwise. Then, in regard to the direction of the arrows, the triangles ∇-shaped have positive orientation and the ones Δ-shaped are negatively orientated. With this notation, the hexagonal tonal cell can be drawn as on the right side of Figure 4.4.

For the general case of a chord extension $[A, B, C]$, two mutually inverted modes $\mu = [A, B, C]$ and $\overline{\mu} = [C, B, A]$ belong to different mode classes. Their shifts have obviously the same property. Then, by

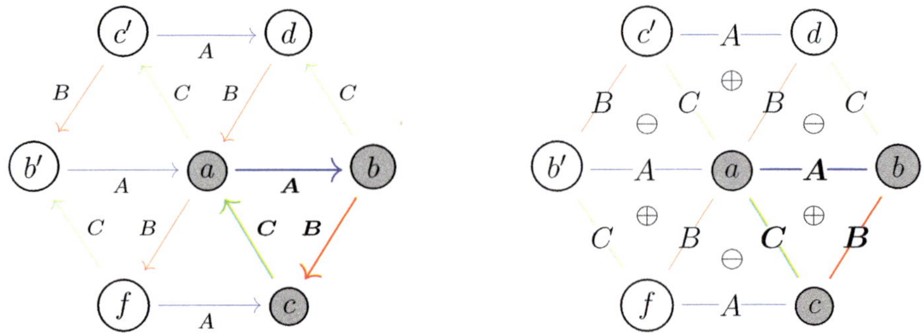

Figure 4.4 (Left) Hexagonal tonal cell of the trichords originated by the directed chord $a\,|\,[A, B, C]$, in boldface. The arrows of the edges indicate how the chords are generated. (Right) The same tonal cell, now assuming the clockwise orientation of the triangles as the positive direction.

taking into account the mode shifts defined in Eq. 2.21 and the definition of inverted modes of the previous section, the family of the symmetric modes of μ can be written as

$$\begin{aligned} \mu &= [A, B, C]; \quad \overline{\mu} = [C, B, A] \\ \mu^1 &= [B, C, A]; \quad \overline{\mu^1} = [A, C, B] \\ \mu^2 &= [C, A, B]; \quad \overline{\mu^2} = [B, A, C] \end{aligned} \tag{4.6}$$

so that each column contains the shifts of a single mode class, with the zero rotations noted as $\mu^0 \equiv \mu$. Let us remark that, for $i \neq 0$, $\overline{\mu^i} \neq (\overline{\mu})^i$, since we had only defined shifts for clockwise directed modes.

With this notation, in Figure 4.5 we are able to draw the diagrams of Figure 4.4 in a simpler form. Nevertheless, since each trichord has three rotations, the description of a chord is not unique. We adopt the criterion of writing the chords shaped as ∇ starting from the note of the *upper-left corner* of the triangle, and the ones shaped as Δ starting with the note of the *lower-right corner*, always clockwise directed. Hence, all the modes appearing in the hexagonal tonal cell are either μ or its inverted mode $\overline{\mu}$, each one representative of a different mode class. These diagrams illustrate the neighbor properties of chords, modes, and notes induced by a chord $a\,|\,[A, B, C]$. They may be interpreted as the topography of

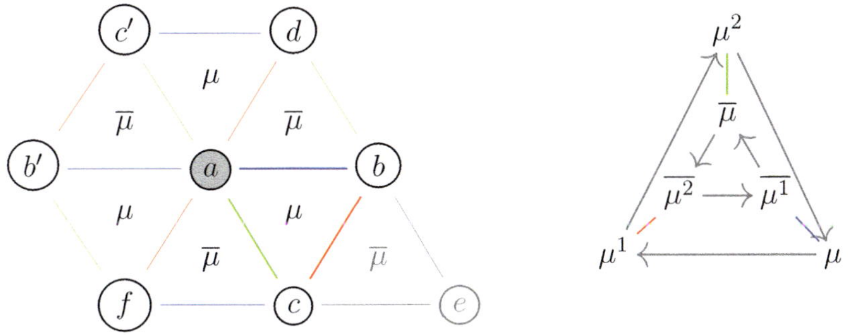

Figure 4.5 (Left) Hexagonal tonal cell [A,B,C] by assuming clockwise as the default orientation of triangles. Each mode is applied to the upper-left corner note if the triangle is ∇-shaped, and to the lower-right corner if it is Δ-shaped. (Right) Cayley graph of $\mathcal{S}_3$ expressing the relationship between modes. A triangular diagram connects each mode class, composed of shifted symmetric modes. Modes of a same orbit are in the vertices, related through a transposition, connected with the color of the interval they share.

a trichord around the root a with regard to its mode, by providing an easy representation of the number of notes shared among chords.

Out of the tonal cell centered in a, the only chord sharing two notes with the initial chord (marked in boldface) is $\{b, c, e\}$, which was placed in the lower-right vertex of the extension. Therefore, this is the closest chord to the initial chord in the path toward another cell not containing the root a. From this viewpoint, that direction indicates the smoothest voice leading path connecting chords of different tonalities. The note e, that does not belong to the tonal cell of the root, corresponds to the leading tone of the major diatonic scales and to the subtonic of the minor scales.

The subject of voice leading and chord geometry can be deepened from comprehensive studies such as [18], [34], and [2].

Notice that geometrical criteria, which are typically used in neo-Riemannian theory of music, are not based on the harmonic distance between notes or between chords, since we are dealing with an equal tempered scale. Nevertheless, as previously discussed, since there is a similarity between a just intonation system with two generators and the Tonnetz with two generators, if the notes a, b, c, e are associated with the tones $\nu_a, \nu_b, \nu_c, \nu_e$, and the intervals A and $-C$ are associated with two coprime harmonics p, q, then the following relationships are simultaneously satisfied

$$b = a{+}A,\ c = a{-}C,\ e = a{+}A{-}C \iff \nu_b = p\,\nu_a,\ \nu_c = q\,\nu_a,\ \nu_e = pq\,\nu_a$$

Hence, a acts as the lowest common ancestor [12] of b and c, and e acts as the lowest common harmonic of b and c, having the chords $\{a, b, c\}$, $\{b, c, e\}$, and $\{a, b, c, e\}$ the same harmonic dissonance.

4.7 CHORD CELL

The dual network of the network in Figure 4.5 is the one in Figure 4.6, where connections between chords are shown instead of connections between tones. More precisely, each triangle representing a chord in the former graph is now a vertex of the latter, and the edges of the triangles in the first graph become vertices in the second one. On the other hand, one note in Figure 4.4 becomes the center of the six vertices of one hexagon in Figure 4.6, which are the chords sharing this note. We call this hexagon *cell of chords*, by labeling it with the central note. In Figure 4.6 all the chords are expressed in terms of the same shift of a mode applied to the same root. The vertices with black circles are chords generated by positive modes, while the chords in white circles, in opposite sides of the diagonals passing through the center, are generated by negative modes.

The whole chord cell with root a can be transported to the same or another cell by applying a *translation*[1] of the root by a single interval A, B, and C (or its complementary $n - A$, $n - B$, and $n - C$) according to the directions of the arrows in the graph. Since each vertex belongs to three different chord cells, the translation of a chord may lead either to contiguous cells or to the same cell. For example, according to Figure 4.6, in the same cell the chords related by a translation are

$$\begin{aligned} \{a + A \,|\, \mu^1\} &= \{a \,|\, \mu^0\} \\ \{a + B \,|\, \mu^2\} &= \{a \,|\, \mu^1\} \\ \{a + C \,|\, \mu^0\} &= \{a \,|\, \mu^2\} \end{aligned} \tag{4.7}$$

[1] Musicians call transpositions to translations, however this may be confusing since in mathematics a transposition is a permutation which exchanges two elements and keeps all others fixed.

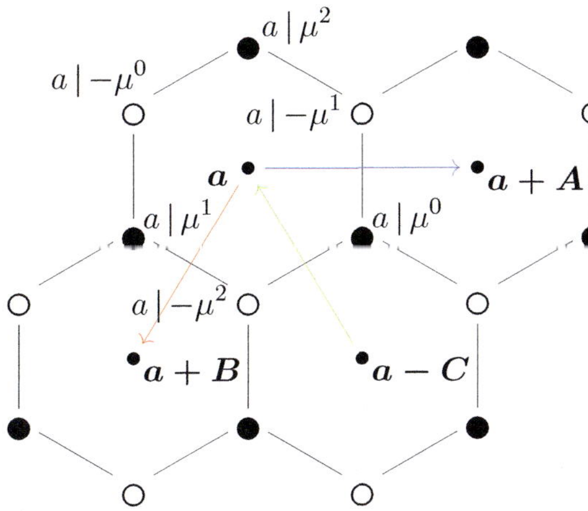

Figure 4.6 Chord cells composed of six chords, half obtained from directed chords with positive modes (black vertices), and half with negative modes (white circles) in opposite vertices of the cell. Similar vertices in contiguous cells are related by translations of the root (in the center of each cell, in boldface) according to the arrows corresponding to intervals of A (blue), B (red), and C (green) units.

$$
\begin{aligned}
\{a+A\,|-\mu^0\} &= \{a\,|-\mu^1\} \\
\{a+B\,|-\mu^1\} &= \{a\,|-\mu^2\} \\
\{a+C\,|-\mu^2\} &= \{a\,|-\mu^0\}
\end{aligned}
\tag{4.8}
$$

with the addition defined in the group $\mathbb{Z}_n$.

In general, the translations along chord cells of the chord $\{x\,|\,\lambda\}$ are defined by successive transformations of the root, according to

$$
\tau_u x = x + u \mod n; \quad u = \pm A, \pm B, \pm C \mod n \tag{4.9}
$$

that preserve the mode. In addition, the new chord maintains the same relative position in the new cells. These transformations acting on the root have the structure of an abelian group with $\tau_{u+v} = \tau_u\tau_v = \tau_v\tau_u$ and identity element $\epsilon = \tau_0$.

Thus, all the chords in the network originated by the directed chord $a\,|\,\mu$ can be expressed from two kinds of transformations over the chord elements:

- Translations τ_u of the root in any of the two independent directions, by preserving the mode.
- Inversion of the mode, by preserving the root of the chord.

It is interesting to point out the cyclic relation of the chords *in the same cell*, by groups of three. For the three chords with positive modes of Eq. 4.7, the relationships expressed from translations on the root can also be written as the action of shifting forward the mode, so that

$$\begin{aligned} \{\tau_{-A}\, a \,|\, \mu^0\} &= \{a \,|\, \mu^1\} \\ \{\tau_{-B}\, a \,|\, \mu^1\} &= \{a \,|\, \mu^2\} \\ \{\tau_{-C}\, a \,|\, \mu^2\} &= \{a \,|\, \mu^0\} \end{aligned} \tag{4.10}$$

For the three chords with negative modes of Eq. 4.8, the translations can be expressed as backward shifts as follows

$$\begin{aligned} \{\tau_{A}\, a \,|\, -\mu^0\} &= \{a \,|\, -\mu^1\} \\ \{\tau_{B}\, a \,|\, -\mu^1\} &= \{a \,|\, -\mu^2\} \\ \{\tau_{C}\, a \,|\, -\mu^2\} &= \{a \,|\, -\mu^0\} \end{aligned} \tag{4.11}$$

4.8 TONAL CELLS [A,B,A] AND [A,2A,A]

We describe the tonal cells for the two remaining particular, degenerate cases of chord extensions. The tonal cell around the note a generated by the mode $[A, B, A]$ is obtained by completing the graph in Figure 4.2. The symmetric modes are

$$\begin{aligned} \mu &= [A, B, A]; & \overline{\mu} &= [A, B, A] = \mu \\ \mu^1 &= [B, A, A]; & \overline{\mu^1} &= [A, A, B] = \mu^2 \\ \mu^2 &= [A, A, B]; & \overline{\mu^2} &= [B, A, A] = \mu^1 \end{aligned} \tag{4.12}$$

with all the modes belonging to a single mode class. The tonal cell is displayed on the left side of Figure 4.7, by using the simplified diagram. The Tonnetz is periodic and symmetric on both sides of the red diagonal through a, and the network may continue toward a number of different cells along this diagonal depending on the interval value B. Notice that, if $g \neq c$, there is no chord not containing the root a that shares two notes with the initial chord, since the lower-right corner of the chord extension

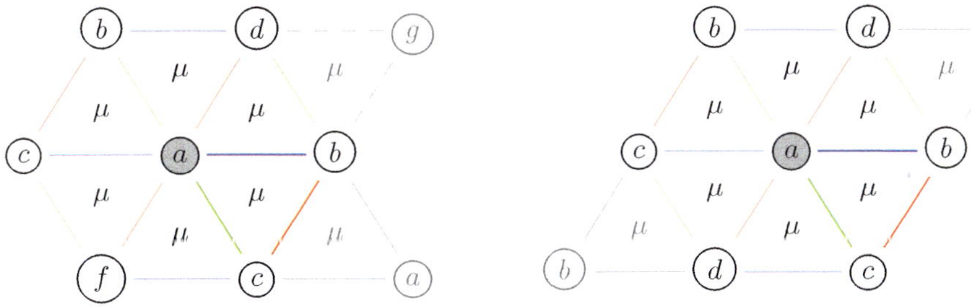

Figure 4.7 (Left) Simplified diagram of the tonal cell centered in a for the Tonnetz $[A, B, A]$. (Right) Tonnetz $[A, 2A, A]$. The full network is depicted, which is periodic and symmetric along and on both sides of the red diagonal through a.

is also the initial chord. Along that direction, the initial chord is always repeated. The case $g = c$ is analyzed in the particular case $[A, 2A, A]$ below.

Therefore, the path toward another tonality must be done through the chords $d \,|\, \mu$ and $f \,|\, \mu$, that only share one note (which is not the root) with the initial chord. Thus, the change toward another tonality is not as smooth as for the $[A, B, C]$ extension.

On the right side of Figure 4.7 the tonal cell for the particular case $2B = n$, i.e., $c = g$, is drawn. It corresponds to the chord extension $[A, 2A, A]$. In this case, we are able to draw the complete network with only one chord $d \,|\, \mu$ non containing the root note a, in the upper-right and lower-lefts corners of the graph. This chord, however, has two notes in common with the initial chord.

4.9 MAJOR AND MINOR CHORDS IN A 12-TET SCALE

Let us see a well-known example of Tonnetz, the one associated with the chord $\{0, 4, 7\} \sim 0 \,|\, [4, 3, 5]$, the major trichord on the fundamental pitch. This tonal network is composed of all the major chords of the 12-TET system generated by the mode $[4, 3, 5]$, as well as all the minor chords generated by the mode $[5, 3, 4]$, as explained in Section 3.5. It is build by beginning at the note 0 and, since $\gcd(4, 3, 5) = 1$, all the notes of the scale appear in this Tonnetz, so that it could be built by beginning at any note. This is a necessary example, since the notes composing the

12-TET scale are obtained as a decomposition into intervals of major and minor thirds of 4 and 3 units.

We first review its relationship in a just intonation system. The FCs, taking values in Ω_0, according to Section 1.1 may be notated by using projection functions, which for a tone α they produce its FC π_α and operate as a commutative group for multiplication. According to this notation, the class of the fundamental is π_1, which in this example we associate with the note C. The major trichord on the fundamental is $M_1 \equiv \{\pi_1, \pi_5, \pi_3\}$, i.e., the C major chord. Its inverted chord, $M_1^{-1} \equiv \{\pi_1, \pi_{\frac{1}{5}}, \pi_{\frac{1}{3}}\}$, is also the F minor chord $m_{\frac{1}{3}}$, and the C minor chord is $m_1 \equiv \pi_3 M_1^{-1} \equiv \{\pi_3\pi_1, \pi_3\pi_{\frac{1}{5}}, \pi_3\pi_{\frac{1}{3}}\} = \{\pi_3, \pi_{\frac{3}{5}}, \pi_1\}$. The other major and minor chords, which are close to the ones of the 12-TET scale, are obtained by applying the projection functions π_3 (the fifth interval) and π_5 (the major third), and its inverses, to the chords M_1 and m_1, respectively.

Let us note that the 12-TET scale, according to its unique decomposition, can be generated from projection functions associated with major and minor thirds, i.e., ${\pi^p}_{2^{\frac{1}{12}}} = {\pi^q}_{2^{\frac{4}{12}}} {\pi^r}_{2^{\frac{3}{12}}}$; $q = 1, 2, 3$; $r = 1, 2, 3, 4$. The major third satisfies $\pi_{2^{\frac{4}{12}}} \approx \pi_5$ and the minor third $\pi_{2^{\frac{3}{12}}} \approx \pi_3\pi_5^{-1}$. Therefore, the chords $\pi_5^{q-r}\pi_3^r M_1$ and $\pi_5^{q-r}\pi_3^r m_1$ are the major and minor chords of the just intonation scale, which are close to those of the 12-TET scale.

Thus, in a 12-TET system the notes are the equally tempered PCs in a $\log_2$ scale, taking values from 0 to 11 along the circle $\mathcal{O}_{12}$. The major trichord of the fundamental C is represented as $0\,|\,[4, 3, 5]$, its inverted chord is $-(0\,|\,[4, 3, 5]) = 0\,|\,[5, 3, 4]$, and the minor chord of the fundamental is $0\,|\,[3, 4, 5]$. To increase a note by an interval of a major third is equivalent to applying a translation τ_4, which is similar to apply the projection function $\pi_5 \approx \pi_{2^{\frac{4}{12}}}$ to the corresponding FC. The opposite translation, $\tau_{-4} \equiv \tau_8$, is equivalent to applying the inverse projection function $\pi_5^{-1} = \pi_{\frac{1}{5}}$. Similarly, a translation τ_7 increases the note by a fifth, which is equivalent to applying the projection function $\pi_3 \approx \pi_{2^{\frac{7}{12}}}$ to the FC, and the opposite translation $\tau_{-7} \equiv \tau_5$ is equivalent to the inverse projection function $\pi_3^{-1} = \pi_{\frac{1}{3}}$, i.e., a fourth. All the major and minor trichords are generated from the symmetric modes of $[4, 3, 5]$. The mode class of $[4, 3, 5]$ generates the major chords and its inverted mode $[5, 3, 4]$ generates the minor ones.

In Figure 4.8 the Tonnetz containing all the major and minor chords of the 12-TET system is represented by explicitly writing the name of

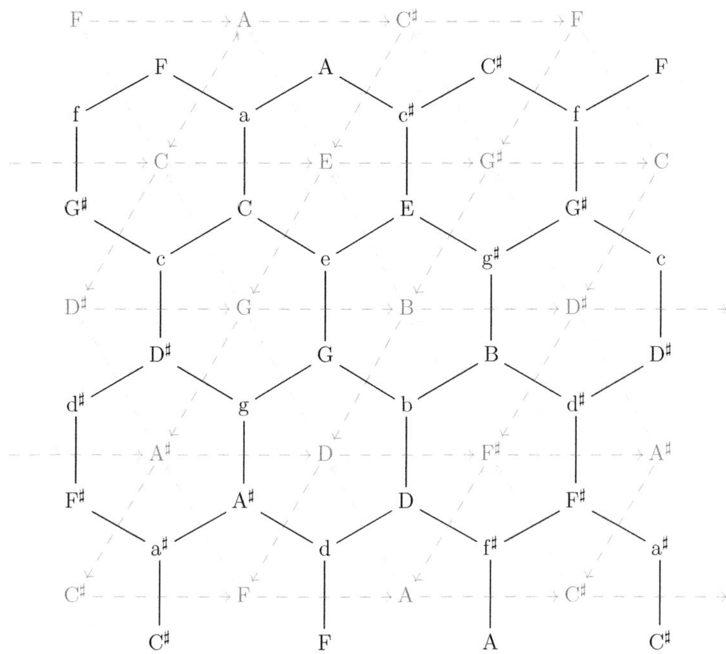

Figure 4.8 Chord cells and Tonnetz from the chord $\{0 \mid [4, 3, 5]\}$ in a 12-TET system. The symmetric modes generate the major (uppercase letters) and minor (lowercase letters) trichords, which are the vertices of the hexagonal chord cells (in black). The chords are surrounded by triangles, which vertices are the tones they contain (in gray).

the notes. This is a double diagram that displays both the tonal network, with notes in gray, and the chord cells, with chords in black, uppercase letters for the major triads and lowercase for the minor ones. Blue, red, and green arrows indicate translations of 4 (a major third), 3 (a minor third), and 5 (a fourth) units, i.e., semitones, respectively. The inverse green arrows correspond to a translation of 7 (a fifth) notes. Each chord is the center of the triangle formed by the notes it contains, and each note is the center of the hexagonal cell of chords containing it.

The connected trichords share two notes. The non-shared notes of two connected trichords differ in 2 semitones for chords vertically connected and in 1 semitone for all the other chords.

A closed path in the Tonnetz along non vertically connected chords is known as a *maximally smooth cycle* [10]. It is straightforward seeing from Figure 4.8 which notes are involved in each maximally smooth cycles for major and minor chords, and which are the common notes among these cycles.

Other properties may be easily derived from the graph. For instance, by starting a continuous polygonal path from any chord in the Tonnetz, we are able to see which chords can be reached by changing only one note in each step.

If the number of steps or edges between two chords is taken as the length of a path over the Tonnetz, then the length of a loop, by beginning and ending in the same chord without repeating any chord, is 6. There are four different maximally smooth cycles of length 6, along the non-vertical polygonal lines.

The minimum length between two chords defines a distance on the Tonnetz. In this example, any two chords are connected by a path of at most of 5 steps. For example, the chord more distant to the F major chord is the d♯ minor chord, at 5 steps distance.

These and other properties of the Tonnetz, such as paths connecting all the chords without repeating any one, loops that may be contracted to a point or may not, algorithms for generating sequences of different chords, etc. can be further studied with the help of an algebra of chords, which is introduced in the following chapters.

CHAPTER 5

Higher-dimensional chords

5.1 HIGHER-DIMENSIONAL TONE NETWORK

The tonal network associated with a trichord $\{a\} = \{a_0 \,|\, \mu\} \in \mathcal{K}(n, 3)$ and the chord network it generates can be generalized to chords and modes of any dimension $k \in \mathbb{N}$. For $k = 3$ the symmetric modes $\mathcal{S}(\mu)$ were expressed as shifts of at most two mode classes, μ and $\overline{\mu}$, but for higher dimensions they contain shifts of several mode classes, obtained from several transpositions of the mode intervals.

By generalizing Eq. 4.2, the tonal network associated with a k-chord $\{a_0 \,|\, \mu\}$, $\mu = [A_0, \ldots, A_{k-1}] \in M(n, k)$, is composed of all the notes in the form $\tau_{c \cdot \mu}\, a_0$, according to the dot product

$$c \cdot \mu = \sum_{i=0}^{k-2} c_i\, A_i \mod n; \quad c = (c_0, \ldots, c_{k-2}) \in \mathbb{Z}^k \tag{5.1}$$

where the sum is extended to a subset of $k - 1$ mode intervals, since one of them is complementary to the others in $\mathbb{Z}_n$. At any node b_0 of the Tonnetz, all the chords having the note b_0 can be generated as a directed chord $b_0 \,|\, \nu$, with $\nu \in \mathcal{S}(\mu)$. Therefore, the *tonal cell* centered at b_0 is composed of the notes forming the $k!$ directed chords generated by the symmetric modes of μ applied to b_0. The chords around b_0 form *chord cell* associated with this root in the dual network of the Tonnetz.

Let us see such a structure in more detail. As in the case described by Eq. 4.1, the Tonnetz associated with a mode with non-two equal intervals can be formally interpreted as a set of points scattered on a

DOI: 10.1201/9781003597964-5

k-dimensional torus. At first, the notes may be represented as an unbounded $(k-1)$-dimensional array of points according to the coordinates given by the intervals of the submode

$$\mu' = [A_0, \ldots, A_{k-2}] \in M(n, k-1)$$

since the interval $A_{k-1} = -(A_0 + \cdots + A_{k-2}) \bmod n$ is uninformative. The intervals are then associated with independent *main directions* of the array, so that we may speak of *vector intervals*. We shall call the direction associated with the last mode interval of a directed chord the *returning direction*. Notice that the returning direction A_{k-1} is independent from any single main direction A_i, $i \neq k-1$. Similarly, it is independent of any subset of different intervals A_i, A_j; $i, j \neq k-1$; up to a subset formed by $k-2$ non-equal intervals without including the returning direction.

By placing a note a_0 at the origin of the array, any point of the array with coordinates $(p_0A_0, \ldots, p_{k-2}A_{k-2}) \bmod n$, with $p_0, \ldots, p_{k-2} \in \mathbb{Z}_n$, represents a note of the n-TET system, of value

$$m = a_0 + p_0A_0 + \cdots + p_{k-2}A_{k-2} \bmod n \tag{5.2}$$

Since the structure is periodic in all directions, the array can be shaped as points over a k-dimensional torus, which is as a $(k-1)$-dimensional hyper surface within $\mathbb{R}^k$. In degenerate cases, where the generating mode has similar intervals, we shall see that by identifying the similar notes on the torus, the Tonnetz may be converted into a band, a Möbius band, a circle, etc.

Nevertheless, according to [35] the Tonnetz may be organized in other ways other than a generalized torus, depending on the properties to study.

Similarly as in Section 4.1, all the notes of the n-TET scale appear on the Tonnetz when $\gcd(A_0, \ldots, A_{k-1}) = 1$. In general, if $\gcd(A_0, \ldots, A_{k-1}) = d$, then d is the smallest positive integer that can be expressed as the combination of Eq. 5.2, which provides a number of $v = \frac{n}{d}$ different notes and vertices of the Tonnetz.

A subset of vertices of the Tonnetz represents a directed k-chord only when they are connected according to a permutation of the $k-1$ intervals of the submode μ', which is equivalent to a permutation of the k intervals of μ. This is because a chord contains k different notes, which can be drawn in the circle of the octave by closing the cycle with the interval A_{k-1} not used in the mode μ'. In addition, over the Tonnetz we may connect the vertices by using all the mode intervals, in the

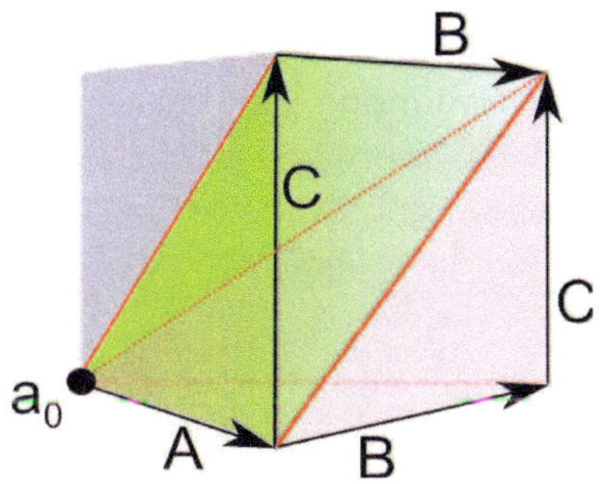

Figure 5.1 Two tetrachords with root a_0.

positive or the negative directions all of them, which is tantamount to reading the mode clockwise or anticlockwise over the octave, although we cannot mix positive and negative intervals. The last and first vertices of every directed chord must also be connected in order to close the octave. This forms a simplex, which is the convex hull of its vertices. Similarly, its subchords are obtained by connecting the vertices along the corresponding returning directions to form the faces. These new edges are linear combination of the independent directions associated with the intervals of μ'.

In Figure 5.1 we show an example for tetrachords, where the two directed chords $a_0 \,|\, [A, B, C, D] = a_0 \,|\, [A, C, B, *]$ and $a_0 \,|\, [A, C, B, D] = a_0 \,|\, [A, C, B, *]$ are drawn. They are two tetrahedra sharing a face. If the intervals of the mode $\mu' = [A, B, C]$ are drawn with equal length in the respective axes, the three main directions are those associated with the edges and sides of a regular cube, while the diagonal lines, marked in red and corresponding to the returning directions of chords and subchords, are linear combinations of them.

5.2 MODULAR STRUCTURE

In the previous section the tonal network has been generalized nearly straightforward from modes in $M(n, 3)$ to $M(n, k)$. Let us do it now in a more rigorous way. The Tonnetz generated by the mode $\mu = [A_0, \ldots, A_{k-1}] \in M(n, k)$ can be viewed as a set of vertices consisting in $(k-1)$-tuples of integers modulo n generated from an initial point $[0, \ldots, 0]$. Therefore, its elements belong to ${\mathbb{Z}_n}^{k-1}$. Since $\mathbb{Z}_n$ is a ring, the Tonnetz has the structure of a $\mathbb{Z}_n$-module with the usual definitions of addition and scalar multiplication as in Euclidean space[1]. Under this viewpoint, the $k-1$ positive mode intervals corresponding to the main directions should be strictly expressed as the $(k-1)$-tuples:

$$\begin{aligned}\vec{A}_0 &= [A_0, 0, \ldots, 0] \\ \vec{A}_1 &= [0, A_1, \ldots, 0] \\ &\cdots\cdots\cdots \\ \vec{A}_{k-2} &= [0, \ldots, A_{k-2}]\end{aligned} \tag{5.3}$$

and the returning direction as

$$\vec{A}_{k-1} = [-A_0, -A_1, \ldots, -A_{k-2}] = -\vec{A}_0 - \vec{A}_1 \cdots - \vec{A}_{k-2}$$

They may be referred to as *vector intervals*, which are the edges or 1-faces of the Tonnetz.

From an initial root, that we may assume as $\vec{a}_0 = [0, \ldots, 0] \in {\mathbb{Z}_n}^{k-1}$ without loss of generality, the Tonnetz is built as the set of vertices that, referred to this root, have vector coordinates given by

$$\vec{m} = p_0\, \vec{A}_0 + \cdots + p_{k-2}\, \vec{A}_{k-2}; \quad p_0, \ldots, p_{k-2} \in \mathbb{Z}_n \tag{5.4}$$

where the addition and product are those of $\mathbb{Z}_n$.

Since the returning direction $\vec{A}_{k-1}$ is independent of any subset of vector intervals containing up to $k-2$ main directions, it may replace any one of the main directions.

Alternatively, any vertex of the Tonnetz may be expressed as

$$\vec{m} = c_0\, \vec{A}_0 + \cdots + c_{k-2}\, \vec{A}_{k-2} + c_{k-1}\, \vec{A}_{k-1}; \quad c_0, \ldots, c_{k-1} \in \mathbb{Z}$$

according to the structure of a module over $\mathbb{Z}$ instead of $\mathbb{Z}_n$.

Each vertex of the Tonnetz is associated with one note, representing an *interval length* from the root. Thus, being the interval lengths $A_i = \ell(\vec{A}_i)$, the interval length corresponding to the vertex $\vec{m}$ is defined as

$$m \equiv \ell(\vec{m}) = p_0\, A_0 + \cdots + p_{k-2}\, A_{k-2} \mod n$$

This generates an equivalence relation, resulting from identifying similar

[1]A module is just a vector space over a ring. In our case, since $\mathbb{Z}_n$ is commutative and it is also the set of scalars, it is no necessary to specify whether the product is defined on the left or on the right.

notes, which induces a structure known as *orbifold.* Thus, the Tonnetz is the object obtained by identifying any two vertices of ${\mathbb{Z}_n}^{k-1}$ which have the same interval length relative to the root.

Therefore, in the Tonnetz, it is satisfied

$$\vec{A}_0 + \cdots + \vec{A}_{k-1} = \vec{0} \tag{5.5}$$

similarly as for the strictly arithmetic relationship, $0 = A_0 + \cdots + A_{k-1} \bmod n$.

5.3 DISTANCE ON THE TONNETZ

Since the length of the vector intervals is A_i for the main directions $i = 0, \ldots, k-2$, and for the returning direction is $\ell(\vec{A}_{k-1}) = A_{k-1} = n - A_0 - \cdots - A_{k-2}$, then the length of an arbitrary path

$$\vec{\Delta} = \sum_{i=1}^{m} p_i \, \vec{C}_i; \quad p_i \in \mathbb{Z}$$

where $\vec{C}_i$ is any one of the vector intervals, is

$$\Delta = \ell(\vec{\Delta}) = \sum_{i=1}^{m} p_i \, \ell(\vec{C}_i) \bmod n$$

Therefore, Δ is a positive integer between 0 and $n-1$ and the length of any path is evaluated with regard to the positive orientation of the vector intervals.

The same structure for the Tonnetz can be built by using negative mode intervals, that is, by following the paths along the opposite orientation. Then, a vector interval $-\vec{A}_i$ has length $n - A_i$.

If $\vec{a}_0$ and $\vec{b}_0$ are two vertices separated by a path $\vec{\Delta}$, then the following expression defines a distance between these vertices,

$$d(\vec{a}_0, \vec{b}_0) = \min\{\Delta, n - \Delta\} = \min\{|a_0 - b_0|, n - |a_0 - b_0|\}$$

that yields the number of tonal units of the n-TET system between a_0 and b_0. In other words, it is the distance between a_0 and b_0 in $\mathbb{Z}_n$.

5.4 NON-DEGENERATE TONNETZ

Two vertices $\vec{a}_0$ and $\vec{b}_0 = \vec{a}_0 + \vec{\Delta}$, with $\vec{\Delta} = \vec{A}_{i_1} + \cdots + \vec{A}_{i_m}$, in the Tonnetz generated by the mode $\mu = [A_0, \ldots, A_{k-1}]$ are connected through

m edges. They can also be connected by paths composed of any permutation of these mode intervals, although there may exist other paths on the Tonnetz connecting these vertices. Other combinations of the mode intervals between $\vec{a}_0$ and $\vec{b}_0 = \vec{a}_0 + \vec{\Delta}'$ must satisfy $\vec{\Delta}' - \vec{\Delta} = \vec{0}$ in ${\mathbb{Z}_n}^{k-1}$ and, therefore, $\Delta' - \Delta = 0 \bmod n$. However, different paths may involve a different number of edges.

In vector notation, a directed k-chord of the Tonnetz with root $\vec{a}_0$ is the set of different vertices involved between $\vec{a}_0$ and $\vec{a}_0 + \vec{A}_{\sigma(0)} + \cdots + \vec{A}_{\sigma(k-1)}$, associated with a permutation $\sigma \in \mathcal{S}_k$, that we notate as

$$\vec{a}_0 \,|\, [\vec{A}_{\sigma(0)}, \ldots, \vec{A}_{\sigma(k-1)}] \tag{5.6}$$

Therefore, the above expression has a *geometric* meaning. The set of interval lengths of the vertices, added to that of the root, is then expressed as $a_0 \,|\, [A_{\sigma(0)}, \ldots, A_{\sigma(k-1)}]$, according to the usual notation, and have an *arithmetic* and tonal meaning.

We say that the Tonnetz generated by a mode μ is *non-degenerate* if its mode intervals are non-two equal. Then the mode classes μ^S do not match. In that case, the notes composing a directed chord such as $a_0 \,|\, [A_0, \ldots, A_{k-1}]$, that is $a_i = a_0 + A_0 + \cdots + A_{i-1}$ with $i = 1, \ldots, k$, are univocally determined either from the latter *arithmetic* expression or from the *geometric* expression $\vec{a}_0 + \vec{A}_0 + \cdots + \vec{A}_{i-1}$. Otherwise, we say that the Tonnetz is degenerate. Then, the same chord may appear in different places over the k-dimensional torus, so that the equivalence relation induced by the distance in $\mathbb{Z}_n$ will associate the Tonnetz with another orbifold, as in the example of Figure 4.3 where each trichord appeared two times.

Hereafter we shall continue by using the arithmetic expression, without vectors, instead of Eq. 5.6, although, according to the context, it will mean the geometric relationship between vertices and edges. If the Tonnetz is non-degenerate, it makes no difference. If it is, it will be pointed out when necessary whether it means vertices and edges or note values and interval lengths[2].

5.5 GENERALIZED TONAL CELL

The notes of the Tonnetz that are vertices of the chords $\{a_0 \,|\, \nu\}$, with $\nu \in \mathcal{S}(\mu)$ and $\mu = [A_0, \ldots, A_{k-1}]$, were referred in Section 4.6 as the

[2]Alternatively, if several main directions have the same interval length A, they can be differentiated by using the notation A, A', A'', and so on, instead of using vectors.

tonal cell around a_0. The number of vertices it contains may be computed as follows. We shall use only positive mode intervals while forming all possible subchords of the network with root a_0.

At one step away from the root, i.e., at the end of one edge, we may reach any vertex expressed as the second vertex of the directed 2-chords in the shape $a_0 \mid [A_i, *]$, i.e., the vertices $a_0 + A_i$. It can be done along the main directions, by choosing i among the indices $0, \ldots, k-2$; and also in the returning direction, with $i = k-1$. Remember that the returning direction is independent from any single main direction. This makes $\binom{k}{1}$ new vertices, in addition to the one corresponding to the root.

At two steps away from a_0 we may reach some new vertices expressed as the third vertex of the directed 3-chords such as $a_0 \mid [A_i, A_j, *]$, with $i \neq j$ among the indices $0, \ldots, k-1$ that are associated with either the main directions or the returning direction. Once again, the path along these intervals follows two independent directions of the network. So, we reach $\binom{k}{2}$ new vertices at two steps away from the root.

This procedure may continue up to the k-th vertex $a_0 + A_{i_1} + \cdots + A_{i_{k-1}}$ of the directed k-chords, for a number of $k-1$ different indices among the values $0, \ldots, k-1$ by providing $\binom{k}{k-1}$ new notes. If one of the intervals is A_{k-1}, we bear in mind that this returning direction is still independent of the other $k-2$ main directions involved the k-chord. This makes a tonal cell with a number

$$N_k = \sum_{i=0}^{k-1} \binom{k}{i} = 2^k - 1 \tag{5.7}$$

of notes, including the root.

5.6 GENERALIZED TONNETZ

According to the preceding section, the geometric structure of a k-chord $\{a\}$ in the Tonnetz is a $(k-1)$-simplex, that is, any note of the chord can be connected with the others in the way described to form a subchord, which is a simplex of lower dimension. The k vertices and notes of the k-chord are the 0-faces of the $(k-1)$-simplex, the $\binom{k}{2} = \frac{k(k-1)}{2}$ edges are the 1-faces forming 2-chords, etc., and the $(k-1)$-face is the whole chord. Hence, each m-face ($m < k$) is an m-simplex and is an $(m+1)$-chord, subchord of $\{a\}$.

By focusing on the vertices of the chord, the following aspects may be pointed out. The $(k-2)$-faces or *facets* of the chord $\{a\}$ are subchords containing $k-1$ notes. If one note not included in $\{a\}$ is added to one facet, we obtain a new k-chord, which is adjacent to $\{a\}$, i.e., shares the $k-1$ vertices of the facet. Therefore, the number of adjacent chords to $\{a\}$ is the same than the number of facets, that is $\binom{k}{k-1} = k$. This set of chords was called the *chord extension* in Section 4.1. In that case, the vertex to be added to the $(k-1)$-subchord is not arbitrary, since these chords belong to the Tonnetz generated by the intervals of μ. As explained in Section 3.7, only two different chords may have a particular subset of $k-1$ vertices in common. They are obtained by writing one chord as the directed chord $a = a_0|[A_0, \ldots, A_{k-3}][A_{k-2}, A_{k-1}]$, so that the directed chord $a' = a_0|[A_0, \ldots, A_{k-3}][A_{k-1}, A_{k-2}]$ is the only one that shares the $k-1$ first vertices of a. By rotations on the original directed chord, we find the k chords sharing the other $k-1$ subsets of vertices.

The chord extension, i.e., the chord $\{a\}$ and its k adjacent chords, may be extended to all the chords over the Tonnetz. The $(k-1)$-simplices associated with the chords form a $(k-1)$-dimensional tessellation of the Tonnetz. The conglomerate of $k!$ simplices, which are the chords sharing one root, is the tonal cell, composed of $N_k = 2^k - 1$ vertices. If the Tonnetz is non-degenerate, these vertices are associated with different notes.

5.7 GENERALIZED CHORD NETWORK

The topography of the chords in the Tonnetz can also be represented with the dual network diagram, the *chord network*, where the chords are associate with vertices instead of the notes. Under this viewpoint, the chords are vertices grouped according to the notes they share, that is, each note is in the center of a chord cell formed by the chords containing it. Each k-chord can then be rooted in anyone of its notes, so that it belongs to k chord cells. In other words, we may think of the k notes composing a k-chord as the center of k adjacent chord cells.

Thus, we now focus in the k-chords of the Tonnetz containing a vertex a_0, i.e., the cell of chords labeled with the root a_0 in its center. It is then possible to write all these chords from directed chords that have a_0 as root. Since the directed chords having the same root can be expressed from the symmetric modes $\mathcal{S}(\mu)$ as $a_0 \,|\, [A_{\sigma(0)}, \ldots, A_{\sigma(k-1)}]$, with $\sigma \in \mathcal{S}_k$, there is a number of $k!$ chords in this cell. Among them, the number of different chords is given by Eq. 3.9.

In general, the chord cell is a *permutahedron* of order k [17], i.e., a $(k-1)$-dimensional polytope embedded in an k-dimensional space, whose

Table 5.1 The geometric structure of the k-chords and its subchords in the tonal network is associated with the simplices (or faces) listed on the left-hand side of the table. The faces of the dual polyhedron on the right-hand side describe the chord cells, associated with roots, that share a number of chords in the chord network. The set of vertices belonging to the k-chords generated around one root is a tonal cell in the tonal network. The set of chords sharing a root in the chord network is the chord cell.

tonal network			*chord network*			
notes	face dimension	object	congruent cells	shared chords	face dimension	object
1	0	vertex	1	k	$k-1$	polyhedron
2	1	edge	2	$k-1$	$k-2$	facet
$\vdots$	$\vdots$	$\vdots$	$\vdots$	$\vdots$	$\vdots$	$\vdots$
$k-1$	$k-2$	facet	$k-1$	2	1	edge
k	$k-1$	polyhedron	k	1	0	vertex
	tonal cell			*chord cell*		
	notes	chords		vertices	facets	
	2^k-1	$k!$		$k!$	2^k-2	

vertices and edges are isomorphic to the Cayley graph of the symmetric group $\mathcal{S}_k$.

Let us note that for a chord $\{x\}$ in the tonal cell a_0, there are chords that belong to its chord extension and other chords that belong to the chord extension of other chords of the cell, which still share the root a_0, although do no share any facet with $\{x\}$. This was illustrated in Figure 4.4, where the trichord with the notes in gray has two adjacent chords in that tonal cell, while the other three chords of the cell do not belong to its chord extension.

To represent the topography of the chords in the Tonnetz we can use the dual network diagram: instead of associating the vertices with notes, we represent the chords by vertices. We shall refer to it as *chord network*. Under this viewpoint, the chords are vertices grouped according to the notes they share, that is, each note is in the center of a chord cell formed by the chords containing it. Each k-chord can then be rooted in anyone of its notes, so that it belongs to k chord cells. In other words, we may think of the k notes composing a k-chord as the center of k adjacent chord cells.

Each couple of chords $\{x\}$ and $\{x'\}$ that have $k-1$ notes[3] in common and differ in one note (hence they share a facet in the tonal network) are adjacent chords of the same cell of chords and are shared by $k-1$ cells of chords. Similarly, a new chord $\{x''\}$, different from $\{x\}$ and $\{x'\}$, that share a facet with $\{x'\}$ in the tonal network (hence differs in one note which does not belong to $\{x\}$, otherwise it would be $\{x''\} = \{x\}$) shares, at least, $k-2$ notes with $\{x\}$ in the chord network. Then, we can connect two chords that have a common facet in the tonal network with an edge in the chord network. Each edge is shared by $k-1$ cells of chords and the cells in the opposite sides of the segment are labeled according to the note they differ.

In general, from the tessellation of the tonal network with vertices as notes and $(k-1)$-simplices as k-chords, we obtain a dual tessellation of the chord network with chords as vertices and notes as polytopes associated with the chord cells. In the former, the $(k-1)$-faces are k-chords, the $(k-2)$-faces are $(k-1)$-subchords, etc., and the 0-faces are notes. In the latter, the 0-faces are k-chords, the 1-faces are $(k-1)$-subchords, etc., and the $(k-1)$-faces are the notes labeling the chord cells. These structures are respectively co-dimensional to $k-1$.

Since in the dual network the notes of the tonal cell become chord cells, the number of $2^k - 2$ vertices around (excluding) the root of the tonal cell becomes the number of

$$f_k = 2^k - 2$$

facets of the chord cell with center at the root. Therefore, each chord cell is surrounded by f_k adjacent chord cells.

5.8 CHORD CELL FACETS

In the chord cell a_0 of the network generated by the directed k-chord $a_0 \mid [A_0, \dots, A_{k-1}]$ with $k > 1$, we pay attention to the chords that share the notes a_0 and $a_0 + A_0$. These chords are generated from directed chords in the form

$$a_0 \mid [A_0][A_{\sigma(1)}, \dots, A_{\sigma(k-1)}]$$

with $\sigma \in \mathcal{S}_{k-1}$. There are $(k-1)!$ of them along different directions of the chord network. They share a facet of the chord cell a_0, which is also

[3]Hence, we shall use the term note to mean a vertex of the tonal network, while the term vertex will be reserved for the chord network.

a facet of the cell $a_0 + A_0$. Since these notes differ in the interval A_0, we shall say that the facet is *orthogonal* to this main direction. Similarly, the chords that share the notes a_0 and $a_0 + A_1$ are generated according to a mode composition

$$a_0 \mid [A_1][A_{\sigma(0)}, \cancel{A_1}, \ldots, A_{\sigma(k-1)}]$$

with $\sigma \in \mathcal{S}_{k-1}$, where the component that is left out in the second partial mode is marked. There are $(k-1)!$ chords that belong to the facet shared by the cells a_0 and $a_0 + A_1$, orthogonal to the main direction A_1.

This may be done for each mode component A_i, until reaching the last facet, which is orthogonal to the returning direction A_{k-1}, shared by the cells a_0 and $a_0 + A_{k-1}$. We call these k disjoint facets *front facets* of the chord cell a_0. By collecting the chords of the k front facets, we get the full chord cell with $k! = k\,(k-1)!$ chords.

For $k > 2$, we may also look for the facets containing the chords that share the two notes a_0 and $a_0 + \Delta$, with $\Delta = A_0 + A_1$. In that case, these chords are generated as either

$$a_0 \mid [A_0, A_1][\cancel{A_0}, \cancel{A_1}, A_{\sigma(2)}, \ldots, A_{\sigma(k-1)}]$$

or

$$a_0 \mid [A_1, A_0][\cancel{A_0}, \cancel{A_1}, A_{\sigma(2)}, \ldots, A_{\sigma(k-1)}]$$

with $\sigma \in \mathcal{S}_{k-2}$, so that there are $2!\,(k-2)!$ of them. This can be done by selecting $\binom{k}{2}$ different couples of intervals for Δ, by determining a similar number of disjoint facets, so that collecting them we get the full set of chords in the cell $k! = \binom{k}{2}\,2!\,(k-2)!$.

Similarly, for $k > m$, the facet containing the chords that share the notes a_0 and $a_0 + \Delta$, with $\Delta = A_0 + \cdots + A_{m-1}$ is composed of chords

$$a_0 \mid [A_{\rho(0)}, \ldots, A_{\rho(m-1)}][\cancel{A_0}, \ldots, \cancel{A_{m-1}}, A_{\sigma(m)}, \ldots, A_{\sigma(k-1)}]$$

with $\rho \in \mathcal{S}_m$ and $\sigma \in \mathcal{S}_{k-m}$, so that there are $m!\,(k-m)!$ of these chords in each facet. In general, we may select $\binom{k}{m}$ subsets of m mode intervals for Δ, which will produce a similar number of disjoint facets. By collecting them, we always get the full set of $k! = \binom{k}{m}\,m!\,(k-m)!$ chords in the cell.

In particular, the case $m = k-1$ leads to facets with a similar number of $(k-1)!$ chords as the front facets. Their last mode interval is a main direction or the returning direction. That is, the notes a_0 and $a_0 + \Delta$ differ in an interval $-A_i$. Hence these facets are also orthogonal to these directions, although they are in the negative part of the axes, by assuming the center of the cell a_0 in the origin. We call them *back facets*. Front and back facets are *principal facets* of the cell of chords. The other facets are the *mid facets*, existing only for $k \geq 4$. The way of moving along the chord network between two chords by maintaining the maximum number of common notes regardless tonality criteria is perpendicular to one facet. Such a movement implies changes between mode classes.

Therefore, once again we obtain the cell of chords as composed of a number of facets

$$f_k = \sum_{m=1}^{k-1} \binom{k}{m} = \binom{k}{1} + \binom{k}{2} + \cdots + \binom{k}{k-2} + \binom{k}{k-1} = 2^k - 2$$

where the facets corresponding to $m = i$ and $m = k - i$, indicated with arrows in the above equation, have the same number $(k-i)!\,i!$ of vertices.

In particular, the number of principal facets in the chord cell is $2k$, corresponding to the first and last terms of the series.

Then, if k is odd, there is a number $\frac{k-1}{2}$ of different chord configurations in the facets, i.e., the amount of different facets with regard to their number of vertices. If k is even, there are $\frac{k}{2}$ different types facets, i.e., with a different number of chords.

Each chord of the cell is contained in each type of facet. We shall see an example for tetrachords in the next section.

5.9 TETRACHORDS

We study in more detail the example of tetrachords shown in Figure 5.1. A tetrachord $\{a\} = \{a_0 \,|\, [A_0, A_1, A_2, A_3]\}$ may be represented by a tetrahedron, i.e., a simplex of dimension 3. Each one of the four faces (facets, in higher dimensions) is a triangle, i.e., a simplex of dimension 2, representing a trichord contained in $\{a\}$. The six edges are simplices of dimension 1, forming subchords of two notes. The vertices are simplex of dimension 0, corresponding to the notes of the chord.

The neighborhood of a tetrachord is as follows. Each face of $\{a\}$ is a subchord of three notes shared by 2 chords[4]. Thus, each face of $\{a\}$ has one adjacent chord, and $\{a\}$ has four adjacent chords, completing

[4]In Section 3.7 we found that the number of chords sharing a set of m consecutive notes with a k-chord is $(k - m + 1)!$. Three notes of a tetrachord are always consecutive.

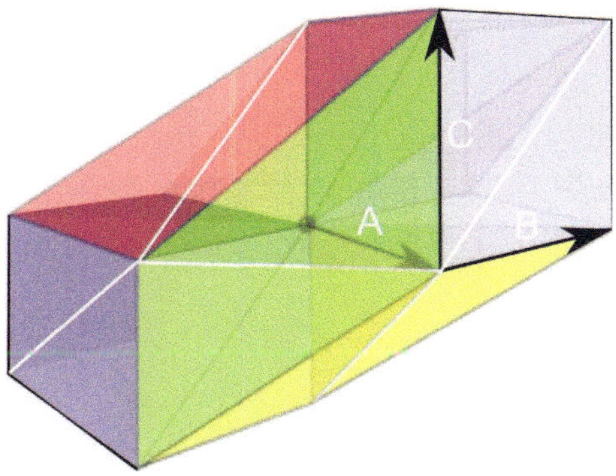

Figure 5.2 Tonal cell of a Tonnetz generated by the mode $[A, B, C, *]$ around a root, represented as a central dot.

the chord extension. Each edge of $\{a\}$ containing two consecutive notes is shared by $3! = 6$ chords and belongs to five neighbor chords different from $\{a\}$. However, the chord $\{a\}$ may also share two non-consecutive notes with other chords. These notes are shared[5] by 4 chords. Finally, a note is shared by $4! = 24$ neighbor chords, composing the chord cell.

The tonal cell is depicted in Figure 5.2. Each tetrachord containing the same root (central black dot) is a tetrahedron; within each one of the two opposite cubes of the figure there are 6 tetrahedra; outside the cubes there are 6 tetrahedra on either side. However, the external faces of Figure 5.2 are shared by couples of adjacent chords. These faces are indeed divided by edges (in white, in the figures) to separate these chords.

The drawing may be better understood with the help of the six orthogonal projections to the main directions $\pm A, \pm B, \pm C$, as shown in Figure 5.3. Each projection gives a hexagonal tonal cell, as we had for trichords. The vertices marked with small circles share four faces, while the other ones are common to six faces. The figure is homotopy equivalent to a regular polyhedron, a *rhombic dodecahedron*, but with the corresponding sides divided by white edges, as shown on the left side of Fig 5.4. Then, by taking into account both colored edges, the above polyhedron becomes a *tetrakis hexahedron*, with 24 faces, 36 edges, and 14 vertices, which in addition to the central root compose the tonal cell.

[5] According to Eq. 2.10, if we express the chord as the i-th rotation in the form $a_i \mid [A_i, A_{i+1}][A_{i+2}, A_{i+3}]$, the shared notes will be the first and last note of the subchord $a_i \mid [A_i, A_{i+1}, *]$. Therefore, these notes are shared by the chords expressed as the former composition, as well as by those generated by the permutations of the intervals in each partial mode.

Figure 5.3 Orthogonal projections to the main axes of the tonal cell of tetrachords. The vertices marked with small circles share four faces and the other ones six faces.

The dual representation for tetrachords, with the chords as vertices, is displayed on the right-hand side of Fig 5.4. The cell of chords centered in the note a_0 is a polyhedron with 24 vertices representing the tetrachords containing this note. Each chord belongs to 4 different chord cells (the own polyhedron, and those sharing the three faces that have the chord as a common vertex) and is the common vertex to 3 edges in each chord cell, joining couples of chords that share three notes. The chords linked by an edge differ in one note. The chords in one face (facet, in higher dimensions) differ in two notes. The chords in a face with 6 vertices differ in two consecutive notes. These are the front and back facets, shown on the left-hand side of Figure 5.5, the former are the colored hexagons and the latter are the opposite non-colored hexagons. The chords in a face with 4 vertices differ in two non-consecutive notes. These are the mid facets, shown on the right-hand side of Figure 5.5 as colored squares. Two adjacent faces have two chords in common.

The resulting polyhedron is a *truncated octahedron*, with 24 vertices, 36 edges, and 14 faces. Indeed, this dual polyhedron is obtained as the dual of the rhombic dodecahedron, a cuboctahedron with

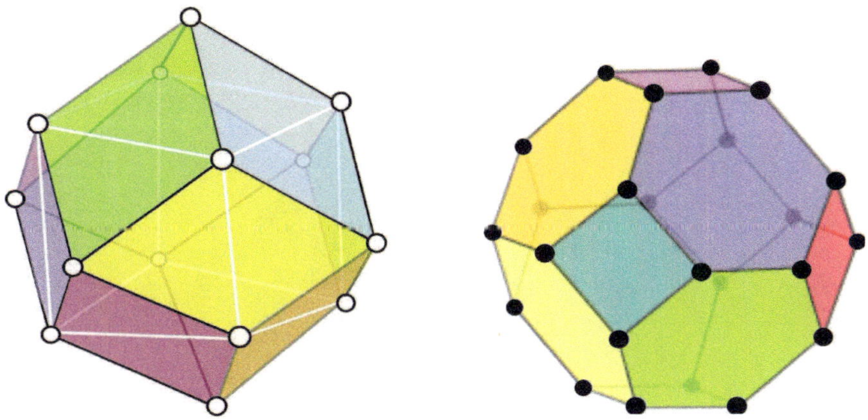

Figure 5.4 The tonal cell of tetrachords is homotopy equivalent to a tetrakis hexahedron (left), where each vertex is a note. It contains $2^4 - 2 = 14$ notes, excluding the root in its center (not drawn). Its dual representation, a chord cell, is a truncated octahedron (right), with chords in the vertices with a common root, in the center of the polyhedron. Each face of the truncated octahedron is a vertex of the tetrakis hexahedron.

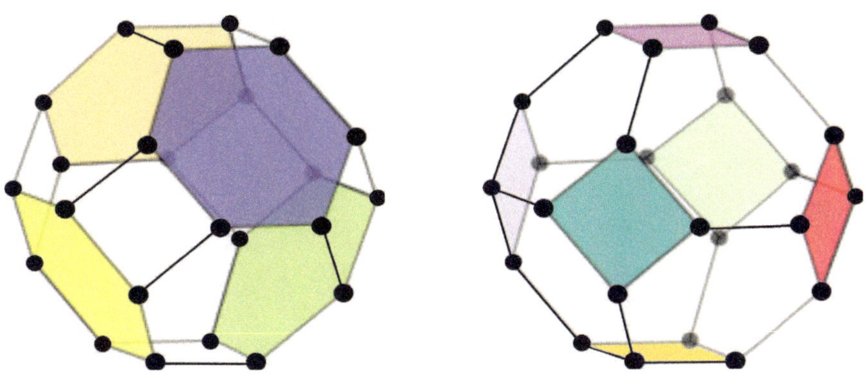

Figure 5.5 (Left) Four front facets (colored hexagons) of the truncated octahedron, with 6 vertices, containing all the chords of the cell. The back facets (non-colored hexagons), opposite to the front facets, also have 6 vertices. (Right) Six mid facets (colored squares) of the cell of chords, with 4 vertices, also containing all the chords of the cell.

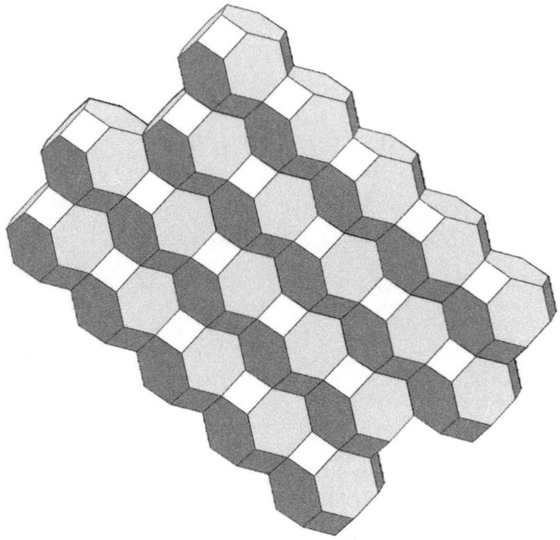

Figure 5.6 Honeycomb of tetrachords tessellating the chord network.

12 vertices, 24 edges, and 14 faces, by alternating 6 squares and 8 triangles, where the vertex of each square is doubled, by transforming the triangles in hexagons. The truncated octahedron tessellates the chord network by translational copies of itself, by forming the honeycomb shown in Figure 5.6.

CHAPTER 6

Operations on the root

6.1 TRANSLATIONS AND INVERSIONS ON DIRECTED CHORDS

It is possible to transform chords over a two- or three-dimensional Tonnetz by using geometrical or graphical means [22,24,28,29], however, for higher dimensional networks it is necessary to define algebraically such transformations. It will be done in the current and following chapters. In such a point, it is a mandatory Lewin's reference the work [26], one of the pioneers in applying group theory to chord transformations. Necessarily with some common notation, the chord operations of the current paper will use a simpler approach, by taking the advantage that we are always moving on the chord network generated by a fixed k-mode.

First, we center our attention in two operations on the root of a directed chord of a n-TET system. They will be noted by Greek letters. The first operation is the *inversion* of the root of a directed chord $a = a_0 \mid \mu$. We define it as

$$\iota(a_0 \mid \mu) = -a_0 \mid \mu$$

Let us recall the equivalence $-a_0 \equiv n - a_0$, since $a_0 \in \mathbb{Z}_n$. If the identity operation on the root is expressed as ϵ, then $\iota^2 = \epsilon$. Hence, the group generated by ι is isomorphic to $\mathbb{Z}_2$. As shown in the last example of Section 3.1, this operation is not defined for chords, since it results in different chords depending on the rotation we use to generate the chord.

The other operation is the *translation* of the root of a directed chord, which induces translations of chords along the tonal network. For $u, v \in \mathbb{Z}_n$, according to the notation

$$\tau_u v = v + u \mod n \tag{6.1}$$

DOI: 10.1201/9781003597964-6

a translation of the directed chord $a_0 \mid \mu \in K(n,k)$ by an interval $u \in \mathbb{Z}_n$ is the transformation

$$\tau_u(a_0 \mid \mu) = \tau_u a_0 \mid \mu \tag{6.2}$$

which is equivalent to a translation of $u \bmod n$ units of the root a_0 in $\mathbb{Z}_n$. In particular, when u matches an interval of the mode μ, we are able to represent geometrically the translation of the directed chord on the Tonnetz.

The Eq. 6.2 allows us to write a directed chord in either of the following equivalent expressions

$$\tau_x \mid \mu \equiv \tau_x(0 \mid \mu) = (x \mid \mu) \tag{6.3}$$

The following equivalences are satisfied

$$\tau_{u+v} = \tau_u \tau_v \tag{6.4}$$

$$\tau_{uv} = \tau_v^u = \tau_v^u \tag{6.5}$$

so that we may write $\tau_u = \tau_1^u$.

Then, the composition of translations has the structure of a cyclic group inherited from the sum in $\mathbb{Z}_n$, generated by the element τ_1. The identity element is $\epsilon = \tau_0 = \tau_1^n$, and the composition satisfies the well-known properties

$$\begin{aligned} \tau_u \tau_v &= \tau_v \tau_u \\ \tau_u(\tau_v \tau_w) &= (\tau_u \tau_v)\tau_w \\ \tau_u \tau_{-u} &= \epsilon \end{aligned} \tag{6.6}$$

Hence, the inverse translation of τ_u may be written in several ways, e.g.,

$$\tau_{-u} = \tau_u^{-1} = \tau_1^{-u}$$

Translations and inversions on the root of a directed chord are non-commutative, $\tau_x \iota \neq \iota \tau_x$. Instead, they satisfy

$$\iota \tau_x = \tau_{-x} \iota \tag{6.7}$$

In summary, in a n-TET system the translations form a group isomorphic to $\mathbb{Z}_n$ and the inversions are a group isomorphic to $\mathbb{Z}_2$. Then, the operations on the root generated by τ_1 and ι have the structure of a semidirect product $\mathbb{Z}_n \rtimes \mathbb{Z}_2$ with $2n$ elements $\{\tau_0, \tau_1, \tau_1^2, \ldots, \tau_1^{n-1}, \iota, \tau_1\iota, \tau_1^2\iota, \ldots, \tau_1^{n-1}\iota\}$, where $\mathbb{Z}_2$ acts on $\mathbb{Z}_n$ by inversion, since the generators of the factor groups satisfy $\iota\, \tau_1\, \iota = \tau_1{}^{-1}$.

Therefore, the operations on the root of directed chords are isomorphic to the dihedral group D_n [19], so called because it is the group of symmetries of a regular polygon.

6.2 TRANSLATIONS ON CHORDS

For a non-null value $u \in \mathbb{Z}_n$, a translation τ_u on a directed chord $a_0 \,|\, \mu$ generates, in general, a different chord, $\{\tau_u a_0 \,|\, \mu\} \neq \{a_0 \,|\, \mu\}$. Nevertheless, when a translation is applied to two directed chords producing the same chord, namely

$$\{a_0 \,|\, \mu\} = \{b_0 \,|\, \nu\}$$

by satisfying the conditions of Eqs. 3.1 or 3.2, then the translated chords are the same, i.e.,

$$\{\tau_u a_0 \,|\, \mu\} = \{\tau_u b_0 \,|\, \nu\}$$

Therefore, we may speak of *translation of a chord* as the result of applying the translation on any rotation of the chord. Hence, we shall write

$$\tau_u\{a_0 \,|\, \mu\} = \{\tau_u a_0 \,|\, \mu\} \tag{6.8}$$

We review some particular translations on chords from those listed in Table 3.1, that involve the mode μ and its inverted modes $-\mu$ and $\overline{\mu}$:

(a) The chords

$$\tau_x\{y \,|\, \mu\} = \tau_y\{x \,|\, \mu\} = \{y + x \,|\, \mu\}$$

and

$$\tau_{-x}\{y \,|\, {-\mu}\} = \tau_y\{-x \,|\, {-\mu}\} = \{y - x \,|\, -\mu\}$$

have a symmetrical distribution about y. For $x = 0$, $\{\tau_y \,|\, \mu\}$ and $\{\tau_y \,|\, {-\mu}\}$ are relative inverted directed chords about y. For $y = 0$, $\{\tau_x \,|\, \mu\}$ and $\{\tau_{-x} \,|\, {-\mu}\}$ are inverted chords.

(b) The translation of the chord $\{x \,|\, \mu\}$ to a new root y may be expressed as

$$\tau_{y-x}\{x \,|\, \mu\} = \tau_y \tau_x^{-1}\{x \,|\, \mu\} = \{y \,|\, \mu\} \tag{6.9}$$

Notice that the translation τ_x^{-1} depends on the root of the chord it is applied to.

(c) In the above equation, if $y = -x$, we get

$$\tau_{-2x}\{x \,|\, \mu\} = \{-x \,|\, \mu\}$$

Hence, the inversion of the root is obtained as a translation also depending on the own root, that can be expressed as

$$\tau_x^{-2}\{x \,|\, \mu\} = \{\iota x \,|\, \mu\} \tag{6.10}$$

6.3 DEPENDENT TRANSLATIONS

The translations of the above examples that depend on the root of the chord or the directed chord to which they are applied may be expressed more clearly without subindices as

$$\tau[a_0]\,(a_0 \,|\, \mu) \equiv \tau_{a_0}\, a_0 \,|\, \mu \tag{6.11}$$

or, in general, for $m \in \mathbb{Z}$,

$$\tau^m[a_0]\,(a_0 \,|\, \mu) \equiv \tau_{a_0}^m\, a_0 \,|\, \mu \tag{6.12}$$

In the same way, we shall write

$$\tau^m[-a_0]\,(a_0 \,|\, \mu) = \tau^{-m}[a_0]\,(a_0 \,|\, \mu) \tag{6.13}$$

We shall also use this notation for arbitrary translations, such as

$$\tau^m[x]\,(a_0 \,|\, \mu) \equiv \tau_x^m\, a_0 \,|\, \mu \tag{6.14}$$

On the other hand, sometimes the translation is carried out by one *interval* of the mode. For example, if the i-th interval of the full mode $\mu \equiv \mu^0 = [A_0, A_1, \ldots, A_{k-1}] \in M(n,k)$ is notated as

$$[\mu]_i = A_i \;; \quad i \in \mathbb{Z}_k \tag{6.15}$$

then we may express a translation of the directed chord $a_0 \,|\, \mu$ by this quantity as

$$\tau[\mu]_i(a_0 \,|\, \mu) \equiv \tau[A_i]\, a_0 \,|\, \mu \tag{6.16}$$

Since the full modes have a cyclic structure, it is possible to refer an interval anticlockwise direction, with a negative index, by defining

$$[\mu]_{-i} \equiv [\mu]_{k-i} = A_{k-i} \;; \quad i \in \mathbb{Z}_k \tag{6.17}$$

A retrograde translation by a mode interval is also possible,

$$\tau[-\mu]_i(a_0 \,|\, \mu) = \tau[-A_i]\, a_0 \,|\, \mu \tag{6.18}$$

where $-\mu$ is the anticlockwise mode $[-A_0, -A_1, \ldots, -A_{k-1}]$. We assume that $\tau[\mu]_0 = \tau_0$.

We shall use the equivalence

$$\tau[-\mu]_i = \tau^{-1}[\mu]_i \tag{6.19}$$

to refer the translation to an interval of the mode μ instead of using its inverted mode. On the other hand, by Eq. 6.17 we can also write the intervals of the positive inversion $\overline{\mu}$ by using the intervals of the mode μ, such as

$$[\overline{\mu}]_i = [\mu]_{-i-1} \tag{6.20}$$

so that we get the equivalence

$$\tau[\overline{\mu}]_i = \tau[\mu]_{-i-1} \tag{6.21}$$

that is valid for positive and negative intervals as well.

In this way, we are able to express *dependent translations* on the root, by the same root or by a mode interval of a directed chord $a_0 \,|\, \mu \in K(n, k)$, with the following operator notation

$$\tau[\cdot](a_0 \,|\, \mu) \equiv \tau[a_0](a_0 \,|\, \mu) \tag{6.22}$$

$$\tau[\cdot]_q(a_0 \,|\, \mu) \equiv \tau[\mu]_q(a_0 \,|\, \mu) = \tau[A_q](a_0 \,|\, \mu) \tag{6.23}$$

With this notation it is necessary to remark some specific behavior of the translations on the root. First, notice that

$$\tau^p[\cdot] \neq (\tau[\cdot])^p \tag{6.24}$$

since each translation on the right-hand side changes the argument of the following translation. Second, instead of satisfying Eq. 6.7, they satisfy

$$\iota\, \tau[\cdot] = \tau[\cdot]\, \iota \tag{6.25}$$

since the argument of the translation on the right-hand side is the inversion of the root, instead of the root.

6.4 PROGRADE TRANSLATIONS BY MODE INTERVALS

Composition of translations according to successive mode intervals are denoted as

$$\begin{aligned} \tau[\mu]_{i,\ldots,i+j} &\equiv \tau[\mu]_i \cdots \tau[\mu]_{i+j} = \\ &= \tau[A_i] \cdots \tau[A_{i+j}] = \tau[A_i + \cdots + A_{i+j}] \end{aligned} \tag{6.26}$$

We pay attention to the root of chord rotations as defined in Eqs. 2.5 and 2.22. The root of the i-th rotation of a directed chord, namely $a^i = a_i \mid \mu^i$, satisfies

$$a_{i+1} = \tau[A_i]a_i = \tau[\mu]_i a_i \tag{6.27}$$

According to Eq. 6.26, the root of successive rotations is

$$a_{i+j} = \tau[A_i + \cdots + A_{i+j-1}]\, a_i = \tau[\mu]_{i,\ldots,i+j-1}\, a_i \tag{6.28}$$

bearing in mind that the whole set of intervals of the full mode satisfy $\tau[A_0 + \cdots + A_{k-1}] = \tau_0$.

In a similar way, translations on the root by a negative interval of the mode satisfy

$$\begin{aligned} a_i &= \tau[-A_i]\, a_{i+1} = \tau[-\mu]_i\, a_{i+1} \\ a_i &= \tau[-(A_i + \cdots + A_{i+j-1})]\, a_{i+j} = \tau[-\mu]_{i,\ldots,i+j-1}\, a_{i+j} \end{aligned} \tag{6.29}$$

by providing the roots of retrograde rotations of the directed chord.

A particular case of Eq. 6.28 is

$$a_i = \tau[A_0 + \cdots + A_{i-1}]\, a_0 = \tau[\mu]_{1,\ldots,i-1}\, a_0 \tag{6.30}$$

where the successive translations begin in the first interval of the mode.

A more compact notation can be used by writing

$$\tau[\mu]_p^q \equiv \tau[\mu]_{p,\ldots,p+q-1} = \tau[A_p + \cdots + A_{p+q-1}] \tag{6.31}$$

where q denotes the number of mode intervals involved in the translation. The foregoing expression is valid for positive and negative intervals p. Notice that the subindex p indicates the lowest mode interval in the composition and the superindex q is the number of compositions.

In general, since $\tau^m[x] = \tau[mx]$ for any $m \in \mathbb{Z}$, it is also hold

$$\tau^m[\mu]_p^q = \tau^m[\mu]_{p,\ldots,p+q-1} \tag{6.32}$$

We notate $\tau[\mu]_p^0 = \tau_0$. It is then satisfied,

$$\tau[\mu]_p^q\, \tau[-\mu]_p^q = \tau_0 \tag{6.33}$$

Also, we are able to write in a shorter form equations 6.28, 6.29, and 6.30. Thus, the *successive notes of a directed chord* satisfy

$$\begin{aligned} a_i &= \tau[\mu]_0^i\, a_0 \\ a_{i+j} &= \tau[\mu]_i^j\, a_i \\ a_i &= \tau[-\mu]_i^j\, a_{i+j} \end{aligned} \tag{6.34}$$

6.5 RETROGRADE TRANSLATIONS BY MODE INTERVALS

We also define successive translations by the mode intervals, read anticlockwise direction,

$$\tau[\mu]_p^{-q} = \tau[\mu]_{p,p-1,\dots,p-q+1} = \tau[A_p + \cdots + A_{p-q+1}] \; ; \quad q \leq p \tag{6.35}$$

which are valid for positive and negative intervals. Here, the subindex p indicates the greatest mode interval in the composition and the superindex q is the number of retrograde compositions. Hence, Eq. 6.33 is also satisfied for negative powers of the translations.

The translations satisfy

$$\tau[\mu]_p^{-q} \, \tau[-\mu]_p^{-q} = \tau_0 \tag{6.36}$$

For example, for $p = -1$ in Eq. 6.35, we may apply translations by the mode intervals starting from the last one, anticlockwise direction, such as

$$\tau[\mu]_{-1}^{-i} = \tau[\mu]_{-1,\dots,-i} = \tau[A_{k-1} + \cdots + A_{k-i}] \tag{6.37}$$

which, by using the positive inversion of the mode, $\overline{\mu} = [A_{k-1}, \dots, A_0]$, may be computed as

$$\tau[\mu]_{-1}^{-i} = \tau[\overline{\mu}]_0^i \tag{6.38}$$

In this way it is possible to express retrograde translations by intervals of a mode as prograde translations by intervals of the positive inversion of the mode.

CHAPTER 7

Operations on the mode

7.1 POSITIVE AND NEGATIVE INVERSIONS

In Section 3.4 we introduced the inverted mode. Now we reformulate this and other operations only involving the mode of a directed chord, without changing the root. These operations will be noted by lowercase Gothic letters.

For a full mode $\mu \equiv \mu^0 = [A_0, A_1, \dots, A_{k-1}] \in M(n,k)$, the negative mode, that is, the anticlockwise form of the inverted mode, is now written as

$$\mathfrak{n}\mu \equiv -\mu = [-A_0, -A_1, \dots, -A_{k-1}] \tag{7.1}$$

and the positive form of the inverted mode is written as

$$\mathfrak{p}\mu \equiv \overline{\mu} = [A_{k-1}, \dots, A_1, A_0] \tag{7.2}$$

Let us recall that in Section 3.4 both forms were used to determine the relative inverted chord of $\{a_0 \,|\, \mu\}$ about a_0.

Thus, the following directed chord, read anticlockwise direction,

$$\mathfrak{n}(a_0 \,|\, \mu) = a_0 \,|\, \mathfrak{n}\mu$$

and the next one, read in the positive orientation,

$$\mathfrak{p}(a_0 \,|\, \mu) = a_0 \,|\, \mathfrak{p}\mu$$

produce the same chord

$$\{a_0 \,|\, \mathfrak{n}\mu\} = \{a_0 \,|\, \mathfrak{p}\mu\} \tag{7.3}$$

Notice that this does not imply that the modes $\mathfrak{n}\mu$ and $\mathfrak{p}\mu$ are the same mode, or that $a_0 \,|\, \mathfrak{n}\mu$ and $a_0 \,|\, \mathfrak{p}\mu$ are the same directed chord.

 DOI: 10.1201/9781003597964-7

For operations on modes, by expressing as $\mathfrak{e}$ the identity transformation, it is obviously satisfied

$$\mathfrak{n}^2 = \mathfrak{p}^2 = \mathfrak{e}$$

7.2 RETROGRADATION AND SHIFTS

In Section 3.2 we referred to the retrograde mode $-\overline{\mu} = -[A_{k-1}, \ldots, A_1, A_0]$, that applied anticlockwise direction to a_0 generates the same chord $\{a_0 \,|\, \mu\} = \{a_0 \,|\, -\overline{\mu}\}$. Thus, we write

$$\mathfrak{r}\mu \equiv \mathfrak{n}\mathfrak{p}\mu = -[A_{k-1}, \ldots, A_1, A_0]; \quad \mathfrak{r}(a_0 \,|\, \mu) = a_0 \,|\, \mathfrak{r}\mu \tag{7.4}$$

Then, we have

$$\{a_0 \,|\, \mathfrak{r}\mu\} = \{a_0 \,|\, \mu\} \tag{7.5}$$

Once more,

$$\mathfrak{r}^2 = \mathfrak{e}$$

It is straightforward to see that

$$\begin{aligned} \mathfrak{r} &= \mathfrak{n}\mathfrak{p} = \mathfrak{p}\mathfrak{n} \\ \mathfrak{n} &= \mathfrak{r}\mathfrak{p} = \mathfrak{p}\mathfrak{r} \\ \mathfrak{p} &= \mathfrak{r}\mathfrak{n} = \mathfrak{n}\mathfrak{r} \end{aligned} \tag{7.6}$$

The set of mode operations $\{\mathfrak{e}, \mathfrak{n}, \mathfrak{p}, \mathfrak{r}\}$ is an abelian Klein four-group, which is the direct product $\langle\mathfrak{n}\rangle \times \langle\mathfrak{p}\rangle$ of cyclic groups of order 2. However, if the modes are used strictly clockwise definite, we only have to deal with $\langle\mathfrak{p}\rangle$.

On the other hand, the shifts on modes defined in Eq. 2.21 can be extended to directed chords as operations that leaves the root unaltered. For a positive mode μ, a prograde single shift is written as

$$\mu^1 \equiv \mathfrak{s}\mu^0 = [A_1, \ldots, A_{k-1}, A_0] \tag{7.7}$$

while, in general, the j-th shift is obtained as

$$\mu^j = \mathfrak{s}\mu^{j-1} = \mathfrak{s}^j\mu^0; \quad j \geq 0 \tag{7.8}$$

with $\mathfrak{s}^0 = \mathfrak{e}$. Obviously, for full modes of dimension k, this transformation is k-periodic, so that $\mathfrak{s}^k = \mathfrak{e}$. Therefore, $\langle\mathfrak{s}\rangle$ is a cyclic group of order k, although the shift does not commute with the previous operations. The exponent of a full mode indicating the number of shift is thus a number in $\mathbb{Z}_k$.

A retrograde or anticlockwise shift may also be applied to μ. That is,

$$\mu^{-1} \equiv \mu^{-1 \bmod k} = \mathfrak{s}^{-1}\mu^0 = [A_{k-1}, A_0, \ldots, A_{k-2}] \tag{7.9}$$

In general, we have

$$\mu^{-j} \equiv \mu^{-j \bmod k} = \mathfrak{s}^{-1}\mu^{-j+1} = \mathfrak{s}^{-j}\mu^0 \tag{7.10}$$

Unlike translations, the operations $\mathfrak{n}, \mathfrak{p}$, and $\mathfrak{s}$ on the modes cannot be extended to chords, since the same operation may result in different chords depending on the directed chord rotation it is applied. They can only be used to transform directed chords with the same root, i.e., *within the same chord cell.* For instance, two rotations of a directed chord, hence producing the same chord, $a_0 \,|\, \mu^0$ and $a_1 \,|\, \mu^1$, after shifting the mode, become $a_0 \,|\, \mu^1$ and $a_1 \,|\, \mu^2 = a_0 + A_0 \,|\, \mu^2$, which are different chords. That is,

$$\begin{gathered} \{a_0 \,|\, \mu^0\} = \{a_1 \,|\, \mu^1\} \\ \{\mathfrak{s}(a_0 \,|\, \mu^0)\} = \{a_0 \,|\, \mu^1\} \neq \{\mathfrak{s}(a_1 \,|\, \mu^1)\} = \{a_1 \,|\, \mu^2\} \end{gathered} \tag{7.11}$$

since a backward rotation of $a_1 \,|\, \mu^2$ is $a_0 + A_0 - A_1 \,|\, \mu^1$, which is a different chord from $a_0 \,|\, \mu^1$, in general.

On the other hand, the retrogradation $\mathfrak{r}$ of a directed chord actually does not modify the mode, but chooses the direction the directed chord is read, without changing the root. Thus, retrogradation can be applied to a chord, although it leaves the chord invariant. Therefore,

$$\mathfrak{r}\{a_0 \,|\, \mu\} = \{a_0 \,|\, \mathfrak{r}\mu\} = \{a_0 \,|\, \mu\} \tag{7.12}$$

is an operation well defined for chords.

7.3 MODE INTERVALS NOTATION

As defined in Eq. 6.15, the intervals $[\mu]_i$ of a mode $\mu = [A_0, A_1, \ldots, A_{k-1}] \in M(n,k)$ exist for $i = \mathbb{Z}_k$. We list some equivalences describing the operations on modes working from mode intervals and rotations $\mu^j, j \in \mathbb{Z}_k$:

$$[\mu]_i = A_i; \quad [\mu^j]_i = A_{i+j} \tag{7.13}$$

Hence, it is satisfied

$$[-\mu^i]_j = [-\mu^{i+j}]_0 \tag{7.14}$$

With regard to the mode operations, we have,

$$\begin{aligned} [\mathfrak{n}\mu]_i &= -A_i; & [\mathfrak{n}\mu^j]_i &= -A_{i+j} \\ [\mathfrak{p}\mu]_i &= A_{-i-1}; & [\mathfrak{p}\mu^j]_i &= A_{-i+j-1} \\ [\mathfrak{r}\mu]_i &= -A_{-i-i}; & [\mathfrak{r}\mu^j]_i &= -A_{-i+j-1} \end{aligned} \tag{7.15}$$

Bearing in mind that $\mathfrak{r} = \mathfrak{n}\mathfrak{p}$, the mode intervals satisfy the following relationships

$$\begin{aligned} [\mathfrak{n}\mu]_i &= [\mathfrak{r}\mu]_{-i-1} = -A_i \\ [\mathfrak{n}\mu]_{-i} &= [\mathfrak{r}\mu]_{i-1} = -A_{-i} \end{aligned} \tag{7.16}$$

With regard to the shifts, we now express equations 7.7 and 7.8 according to the mode intervals. In general, the intervals of a mode shift satisfy

$$[\mathfrak{s}^j\mu]_i = [\mu]_{i+j} \tag{7.17}$$

which is valid for $i, j \in \mathbb{Z}_k$.

7.4 PROPERTIES

By using the foregoing relationships, it is easy to prove the following properties of the mode operations involving shifts, one commutative and two anticommutative:

$$\begin{aligned} \mathfrak{s}\mathfrak{n} &= \mathfrak{n}\mathfrak{s} \\ \mathfrak{s}\mathfrak{p} &= \mathfrak{p}\mathfrak{s}^{-1} \\ \mathfrak{s}\mathfrak{r} &= \mathfrak{r}\mathfrak{s}^{-1} \end{aligned} \tag{7.18}$$

Therefore, for any $j \in \mathbb{Z}$, it as also fulfilled

$$\begin{aligned} & \mathfrak{s}^j\mathfrak{n} = \mathfrak{n}\mathfrak{s}^j \\ \mathfrak{s}^j\mathfrak{p} = \mathfrak{p}\mathfrak{s}^{-j}; \quad & \mathfrak{p}\mathfrak{s}^j = \mathfrak{s}^{-j}\mathfrak{p} \\ \mathfrak{s}^j\mathfrak{r} = \mathfrak{r}\mathfrak{s}^{-j}; \quad & \mathfrak{r}\mathfrak{s}^j = \mathfrak{s}^{-j}\mathfrak{r} \end{aligned} \tag{7.19}$$

We deduce another basic property. Let us recall that the shifts $\mathfrak{s}^j\mu$, $j \in \mathbb{Z}_k$, of the full mode $\mu \in M(n, k)$ determine one mode class μ^S. The inverse mode, e.g., written as a positive mode $\mathfrak{p}\mu$, and their shifts belong also to one mode class $(\mathfrak{p}\mu)^S$, not necessarily the same as μ^S. For the sake of the second relationship of Eq. 7.18, these shifts satisfy

$$\mathfrak{p}\mathfrak{s}^j\mu = \mathfrak{s}^{-j}\mathfrak{p}\mu$$

so that all the inversions of the modes in μ^S belong to the mode class $(\mathfrak{p}\mu)^S$. Therefore, we may speak of mutually inverted mode classes, which have the same cardinality.

In particular, if μ and $\mathfrak{p}\mu$ belong to the same class, i.e., for some $m \in \mathbb{Z}_k$, $\mathfrak{p}\mu = \mathfrak{s}^m\mu$, then $(\mathfrak{p}\mu)^S = \mu^S$, since, for any $j \in \mathbb{Z}_k$, $\mathfrak{p}\mathfrak{s}^j\mu = \mathfrak{s}^{m-j}\mu$.

7.5 NEIGHBOR CHORDS

The way of moving between chords along the chord network by maintaining the maximum number of common notes is along the edges of congruent cells, where, between adjacent vertices only one note is changed. This is a general case that includes parsimonious voice leading [11] along the Tonnetz generated by a major triad and its dual network.

Originally, the concept of parsimonious voice leading was applied when two notes of a trichord were maintained during a transformation and the third note moved by a minor second or a major second. In [18] it was given a more flexible definition by requiring that just one note was maintained, with the other two moving by minor or major second.

In such a path, the chords at the ends of the edge belong to each one of the cells that share this edge, but depending on the rotation we use to describe them, we can make explicit that they also belong to the opposite non-congruent cells that are separated by this edge. In a non-degenerate Tonnetz, this efficient path *combines chords generated by different mode classes.*

Before describing how to move in the chord network from one chord to its neighbor, we must remark that some of the mode transformations can also be applied to partial modes. In particular, the operators $\mathfrak{p}$ and $\mathfrak{s}$ may act on partial modes by maintaining the total mode as a full mode. For instance, given a full mode written as a composition of partial modes, i.e., $\mu = \nu \cdot \kappa$, with $\nu \in M(n,p), \kappa \in M(n,q)$ and $p + q = k$, we can generate new full modes such as $\mathfrak{p}\nu \cdot \mathfrak{s}\kappa$, $\mathfrak{p}\mathfrak{s}\nu \cdot \kappa$, etc. The only condition is not to mix positive and negative partial modes.

For a mode $\mu = [A_0, A_1][A_2, \ldots, A_{k-1}]$, $k > 2$, we notate a single transposition operator as

$$\mathfrak{t}\mu = \mathfrak{p}[A_0, A_1][A_2, \ldots, A_{k-1}] \tag{7.20}$$

It is the transposition of the first two mode intervals. Obviously, $\mathfrak{t}^2 = \mathfrak{e}$.

The ith transposition of two consecutive intervals A_i, A_{i+1} is obtained by combining the foregoing operator with the mode shifts as

$$\mathfrak{t}_i = \mathfrak{s}^{-i}\mathfrak{t}\,\mathfrak{s}^i, \quad i \in \mathbb{Z}_k \tag{7.21}$$

In this way it is possible to reach all the transpositions of the symmetric group $\mathcal{S}_k$. The $k-1$ generators $\{\mathfrak{t}_0, \ldots, \mathfrak{t}_{k-2}\}$ provide a presentation of $\mathcal{S}_k$ $(k > 2)$ with the following constraints

$$\begin{aligned} &\text{(i)} \quad \mathfrak{t}_i^2 = \mathfrak{e} \\ &\text{(ii)} \quad (\mathfrak{t}_j\mathfrak{t}_i)^2 = \mathfrak{e} \Leftrightarrow |i-j| \neq 1 \\ &\text{(iii)} \quad (\mathfrak{t}_{i+1}\mathfrak{t}_i)^3 = \mathfrak{e} \end{aligned} \tag{7.22}$$

Equivalently,

$$\begin{aligned} &\text{(i)} \quad \mathfrak{t}_i^{-1} = \mathfrak{t}_i \\ &\text{(ii)} \quad \mathfrak{t}_j\mathfrak{t}_i = \mathfrak{t}_i\mathfrak{t}_j \Leftrightarrow |i-j| \neq 1 \\ &\text{(iii)} \quad \mathfrak{t}_{i+1}\mathfrak{t}_i\,\mathfrak{t}_{i+1} = \mathfrak{t}_i\mathfrak{t}_{i+1}\mathfrak{t}_i \end{aligned} \tag{7.23}$$

The second condition only applies for $k > 3$.

These constraints are also valid *for all* the indices in $\mathbb{Z}_k$, by assuming that the second condition $|i-j| \neq 1$ takes place in $\mathbb{Z}_k$, i.e., it that does not apply to the pair $\mathfrak{t}_0$ and $\mathfrak{t}_{k-1}$.

7.6 OPERATING RULES FOR TRANSPOSITIONS

Let us see some results involving transpositions.

▷ For $j = 2, \ldots, k$, given a subset of $j-1$ consecutive numbers $\{i, i+1, \ldots, i+j-2\}$ in $\mathbb{Z}_k$ and a permutation $\sigma \in \mathcal{S}_{j-1}$, it is satisfied

$$(\mathfrak{t}_{\sigma(i+j-2)} \cdots \mathfrak{t}_{\sigma(i+1)}\,\mathfrak{t}_{\sigma(i)})^j = \mathfrak{e} \tag{7.24}$$

This is because any composition of $j-1$ different transpositions involving a subset of consecutive indices in $\mathbb{Z}_k$ (regardless the order of composition) is a cycle of length j and is a permutation of order j.

In particular, by assuming the indices in $\mathbb{Z}_k$, this applies to a full mode shift expressed from any composition in the form

$$\mathfrak{s} = \mathfrak{t}_{i+k-2} \cdots \mathfrak{t}_{i+1}\mathfrak{t}_i \tag{7.25}$$

The transposition $\mathfrak{t}_{k-1}$ can be expressed in several ways in terms of the others, for example, since $\mathfrak{s} = \mathfrak{t}_{k-2} \cdots \mathfrak{t}_0$, then

$$\mathfrak{t}_{k-1} = \mathfrak{t}_{k-2} \cdots \mathfrak{t}_1\, \mathfrak{t}_0\, \mathfrak{t}_1 \cdots \mathfrak{t}_{k-2} \tag{7.26}$$

When the single transposition operator $\mathfrak{t} = \mathfrak{t}_0$ is applied to the directed chord $a_0 \,|\, \mu$, the resulting directed chord $\mathfrak{t}(a_0 \,|\, \mu) = a_0 \,|\, \mathfrak{t}\mu = (a_0, a_0 + A_1, a_0 + A_0 + A_1, a_0 + A_0 + A_1 + A_2, \ldots, a_0 + A_0 + \cdots, A_{k-1})$ has exchanged the note $a_1 = a_0 + A_0$ of the original chord for $a_0 + A_1$. Both directed chords are still referred to the same cell, although after one rotation, both chords can be written as

$$\begin{aligned} \{a_0 \,|\, \mu\} &= \{a_0 + A_0 \,|\, \mathfrak{s}\mu\} = \\ &= \{a_0 + A_0, a_0 + A_0 + A_1, \ldots, a_0 + A_0 + \cdots, A_{k-1}, a_0\} \\ \{\mathfrak{t}(a_0 \,|\, \mu)\} &= \{a_0 + A_1 \,|\, \mathfrak{st}\mu\} = \\ &= \{a_0 + A_1, a_0 + A_0 + A_1, \ldots, a_0 + A_0 + \cdots, A_{k-1}, a_0\} \end{aligned} \tag{7.27}$$

now explicitly referred to the different cells centered in the notes they differ. Therefore, the operator $\mathfrak{t}$ has displaced the original chord along the edge connecting the cells $a_1 = a_0 + A_0$ and $a_0 + A_1$.

In general, if the transposition operator is applied to the ith rotation of the directed chord, $a^i = a_i \,|\, \mu^i$, it is immediate to prove the following result.

▷ The chords $\{a_i \,|\, \mu^i\}$ and $\{\mathfrak{t}(a_i \,|\, \mu^i)\}$ only differ in the notes $a_{i+1} = a_i + A_i$ and $a_i + A_{i+1}$, respectively.

The reasoning is similar to that of Eq. 7.27. The chord $\{a_0 \,|\, \mu\}$ is equivalent to its rotations $\{a_i \,|\, \mu^i\} = \{a_i + A_i \,|\, \mathfrak{s}\mu^i\}$, while $\{\mathfrak{t}(a_i \,|\, \mu^i)\} = \{a_i + A_{i+1} \,|\, \mathfrak{st}\mu^i\}$. These roots correspond to opposite cells in the ends of the edge that connects these chords.

To recover the original position of the common notes, it suffices to do the ith inverse rotation. For $i = 0, \ldots, k-2$, the root a_0 appears in the inversions of the directed chords $\mathfrak{t}(a_i \,|\, \mu^i)$, i.e.,

$$\begin{aligned} \{\mathfrak{t}(a_0 \,|\, \mu^0)\} &= \{a_0 \,|\, [A_1, A_0, \ldots, A_{k-1}]\} \\ \{\mathfrak{t}(a_i \,|\, \mu^i)\} &= \{a_0 \,|\, [A_0, \ldots, A_{i+1}, A_i, \ldots]\}; \quad i = 1, \ldots, k-2 \end{aligned} \tag{7.28}$$

For $i = k-1$, the root a_0 is exchanged by the note $a_0 + A_0 - A_{k-1}$, i.e.,

$$\{\mathfrak{t}(a_{k-1} \,|\, \mu^{k-1})\} = \{a_0 + A_0 - A_{k-1} \,|\, [A_{k-1}, A_1, \ldots, A_{k-2}, A_0]\} \tag{7.29}$$

This yields the following result.

▷ For $i = 0, \ldots, k-2$, the chords $\{\mathtt{t}(a_i \mid \mu^i)\}$ remain in the chord cell a_0. The chord $\{\mathtt{t}(a_{k-1} \mid \mu^{k-1})\}$ does not belong to the chord cell a_0. The edge connecting the chord cells a_0 and $a_0 + A_0 - A_{k-1}$ is the one to follow in order to move to a non-congruent cell with a_0.

At the end of Section 8.6, we come back to these transformations in more detail.

7.7 TRANSLATIONS BY MODE INTERVALS

▷ Prograde translations of a mode $\mu \in M(n,k)$ satisfy

$$\tau[\mu^{i-j}]_0^j = \tau\left[\mathfrak{p}\mu^i\right]_0^j \tag{7.30}$$

According to Eq. 2.21, since the subindices of the mode intervals belong to $\mathbb{Z}_k$, we have

$$\mu^i = [A_{i-k}, \ldots, \overbrace{A_{i-j}, \ldots, A_{i-1}}^{j}]$$

$$\mathfrak{p}\mu^i = [\overbrace{A_{i-1}, \ldots, A_{i-j}}^{j}, \ldots, A_{i-k}]$$

Similarly,

$$\mu^{i-j} = [\overbrace{A_{i-j}, A_{i-j+1}, \ldots, A_{i-j+j-1}}^{j}, \ldots, A_{i-j+k-1}]$$

Hence, the sums of the j first terms of both foregoing equations is

$$\tau\left[\mathfrak{p}\mu^i\right]_0^j = \tau[\mu^{i-j}]_0^j = \tau[A_{i-j} + \cdots + A_{i-1}]$$

leading to the result expressed in Eq. 7.30.

In particular, for $i = 0$ we get

$$\tau[\mu^{-j}]_0^j = \tau[\mathfrak{p}\mu^0]_0^j 1 \tag{7.31}$$

and, for $i = j$, we get

$$\tau[\mu^0]_0^j = \tau\left[\mathfrak{p}\mu^j\right]_0^j \tag{7.32}$$

which are particular cases of Eq. 7.30.

7.8 RELATIONSHIPS INVOLVING SHIFTS AND TRANSLATIONS

We give several properties combining translations by intervals of a shifted mode, which will be used in the following sections. Notice that the equations involving translations by mode intervals are also valid for inverse translations, i.e., translations applied to negative mode intervals.

1. An interval A_{i+j} of the mode $\mu = [A_0, A_1, \dots, A_{k-1}]$ can be denoted in different ways, depending on the mode shift and the chosen interval, as follows,

$$[\mu^{i+j}]_0 = [\mu^i]_j = [\mu^j]_i \tag{7.33}$$

   with indices in $\mathbb{Z}_k$. In other words, the indices corresponding to the shift and to the chosen interval compensate each other.

2. According to Eq. 6.38, the j-th shift of the mode can be expressed as

$$\tau[\mu^j]_{-1}^{-i} = \tau\left[\mathfrak{p}\mu^j\right]_0^i \tag{7.34}$$

   that equals $\tau[A_{j+k-1} + \cdots + A_{j+k-i}]$.

3. By taking into account equations 6.31 and 7.33, for the j-th shift of a full mode, the following prograde translation with $1 \leq i \leq k$, begins with the mode interval A_j and has i compositions,

$$\tau[\mu^j]_0^i = \tau[\mu^{j-l}]_l^i \tag{7.35}$$

   for any integer l. Clearly, the above translation is equivalent to $\tau[A_j + \cdots + A_{j+i-1}]$.

4. Since the first i intervals of μ^j are the last i intervals of $\mathfrak{p}\mu^j$, in reverse order, their sum match. Thus, $\tau[\mu^j]_0^i = \tau[\mathfrak{p}\mu^j]_{k-i}^i$. Hence, by equations 7.19, 7.33, and 7.35, we get

$$\tau[\mu^j]_0^i = \tau[\mathfrak{p}\mu^{j+i+l}]_l^i \tag{7.36}$$

5. The sum of the intervals of a full mode is 0 in $\mathbb{Z}_n$, hence $\tau[\mu^j]_0^k = \tau_0$. Thus, $\tau[\mu^j]_0^i\tau[\mu^j]_i^{k-i} = \tau_0$. Therefore, in $\mathbb{Z}_n$ we have

$$\tau[\mu^j]_0^i = \tau[\mathfrak{n}\mu^j]_i^{k-i} \tag{7.37}$$

6. Owing to equations 7.19 and 7.35, going backwards $i - l$ intervals and compensating with a prograde shift of $i - l$ steps, we then get the more general relationship,

$$\tau[\mu^j]_0^i = \tau[\mathfrak{n}\mu^{j+i-l}]_l^{k-i} \tag{7.38}$$

7. We give an alternative version of Eq. 6.34. By taking into account Eq. 7.33, the notes of a directed chord $a_0 \,|\, \mu$ satisfy

$$a_{i+j} = \tau[\mu^i]_0^j \, a_i \tag{7.39}$$

8. By inverting the above equation, we get $a_i = \tau[\mathfrak{n}\mu^i]_0^j \, a_{i+j}$. However, Eq. 7.39 can also be written by taking into account equations 7.35 and 7.37 as $a_{i+j} = \tau[\mathfrak{n}\mu^i]_j^{k-j} \, a_i = \tau[\mathfrak{n}\mu^{i+j}]_0^{k-j} \, a_i$. Hence, by applying the opposite translation and rewriting the indices, we get the retrograding notes as

$$a_{i-j} = \tau[\mu^i]_0^{k-j} \, a_i \tag{7.40}$$

CHAPTER 8

Chord transformations

8.1 OPERATIONS ON ROOT AND MODE

As mentioned above, the operations on the mode $\mathfrak{n}, \mathfrak{p}$, and $\mathfrak{s}$ are defined only for directed chords. According to the condition given by Eq. 3.2 of uniqueness of chords, the operations on two directed chords generating the same chord will produce the same chord only if the operations that modify their modes (with definite orientation) are balanced with operations on their roots.

Inversion of directed chords, either using the positive or the negative mode inversions, combine operations on the root and the mode. Both forms of inversion applied to a directed chord generate the same chord, although read in opposite orientations. The operations on directed chords involving root and mode will be noted by uppercase Gothic letters.

We write the inversion of a directed chord, under operational form and by using the positive inverted mode, such as

$$\mathfrak{I}(a_0 \,|\, \mu) = \iota\mathfrak{p}(a_0 \,|\, \mu) = -a_0 \,|\, \mathfrak{p}\mu \tag{8.1}$$

Similarly, by using the negative inversion of the mode, we write the inversion as the mirror-directed chord

$$\mathfrak{M}(a_0 \,|\, \mu) = \iota\mathfrak{n}(a_0 \,|\, \mu) = -a_0 \,|\, \mathfrak{n}\mu \tag{8.2}$$

We notate the identity transformation of a directed chord as $\mathfrak{E} = \epsilon\,\mathfrak{e}$, which is the commutative composition of both identities acting on the root and the mode. Then, $\mathfrak{I}^2 = \mathfrak{M}^2 = \mathfrak{E}$. Let us remember that, as directed chords, $\mathfrak{I}(a_0 \,|\, \mu) \neq \mathfrak{M}(a_0 \,|\, \mu)$, although they generate the same chord after reordering their notes. In the next section we shall review, as

DOI: 10.1201/9781003597964-8

studied in Section 3.4, that they are well-defined operations on chords. Generally, when dealing with chords, we shall notate the inversion with $\mathfrak{I}$.

According to equations 8.1 and 8.2, and bearing in mind equations 6.7 and 6.25, it is straightforward to derive the following equivalences that combine inversions, translations, and dependent translations,

$$\begin{aligned} \mathfrak{I}\ \tau_x(a_0 \,|\, \mu) = \tau_{-x}\mathfrak{I}(a_0 \,|\, \mu) \Rightarrow \mathfrak{I}\ \tau[\ ](a_0 \,|\, \mu) = \tau[\ ]\ \mathfrak{I}(a_0 \,|\, \mu) \\ \mathfrak{M}\ \tau_x(a_0 \,|\, \mu) = \tau_{-x}\mathfrak{M}(a_0 \,|\, \mu) \Rightarrow \mathfrak{M}\ \tau[\cdot](a_0 \,|\, \mu) = \tau[\cdot]\ \mathfrak{M}(a_0 \,|\, \mu) \end{aligned} \tag{8.3}$$

8.2 INVERSION OF CHORDS

We illustrate the use of the above properties of mode operations by producing a new derivation of Eq. 3.6, meaning that the inversion of a chord is well defined regardless the directed chord we use to define it. A numerical case was also shown in the third example of Section 3.1.

Let us see that, if the directed chords $a_0 \,|\, \mu$ and $b_0 \,|\, \nu$ define the same chord

$$\{a_0 \,|\, \mu\} = \{b_0 \,|\, \nu\}$$

then $\mathfrak{I}(a_0 \,|\, \mu)$ and $\mathfrak{I}(b_0 \,|\, \nu)$ do also define a unique chord

$$\{\mathfrak{I}(a_0 \,|\, \mu)\} = \{\mathfrak{I}(b_0 \,|\, \nu)\}$$

The result is identical in modes oriented anticlockwise, with the operator $\mathfrak{M}$.

According to Eq. 3.2, there is one value $m \in \mathbb{Z}_k$ such that $a_m = b_0$ and $\mu^m = \nu$. Hence, the following directed chords coincide:

$$a_m \,|\, \mathfrak{s}^m \mu = b_0 \,|\, \nu \tag{8.4}$$

Then, for the respective inverted directed chords $\mathfrak{I}(a_0 \,|\, \mu) = -a_0 \,|\, \mathfrak{p}\mu$ and $\mathfrak{I}(b_0 \,|\, \nu) = -b_0 \,|\, \mathfrak{p}\nu$ we look for the index i that satisfies

$$-a_m \,|\, \mathfrak{s}^i \mathfrak{p}\mu = -b_0 \,|\, \mathfrak{p}\nu \tag{8.5}$$

By inverting the directed chords of Eq. 8.4, according to Eq. 8.1 we get

$$-a_m \,|\, \mathfrak{p}\mathfrak{s}^m \mu = -b_0 \,|\, \mathfrak{p}\nu \tag{8.6}$$

which can be expressed, by applying the second relationship of Eq. 7.18, as

$$-a_m \,|\, \mathfrak{s}^{-m} \mathfrak{p}\mu = -b_0 \,|\, \mathfrak{p}\nu \tag{8.7}$$

Then, by comparing Eq. 8.5 and Eq. 8.7, we obtain the shift

$$i = k - m$$

which makes the inverted directed chords fit them together.

It is straightforward to derive the reciprocal result; that is, if the inversion of two directed chords generates the same chord, then both directed chords also generate the same chord, since $\mathfrak{I}^2$ is the identity.

As a consequence of the above property, in the third example of Section 3.1, we got the value $i = 2$.

Therefore, the inversion is well defined for chords and it can be applied to any directed chord of the class. Then, the inversion of a chord is given by either of the following equalities,

$$\mathfrak{I}\{a_0 \,|\, \mu\} = \mathfrak{M}\{a_0 \,|\, \mu\} = -\{a_0 \,|\, \mu\} \tag{8.8}$$

According to Eq. 8.3, inversion of chords satisfies

$$\mathfrak{I}\ \tau_x = \tau_{-x}\mathfrak{I}\ ; \qquad \mathfrak{M}\ \tau_x = \tau_{-x}\mathfrak{M} \tag{8.9}$$

8.3 INVERSION AND MIRROR BY X

We may also define other operations on the root and mode of a directed chord, such as the inversion by an interval or index x, that is,

$$\mathfrak{I}_x(a_0 \,|\, \mu) \equiv \tau_x\, \mathfrak{I}(a_0 \,|\, \mu) = \tau_x\, \iota\, \mathfrak{p}(a_0 \,|\, \mu) = (x - a_0) \,|\, \mathfrak{p}\mu \tag{8.10}$$

or, in the mirror form,

$$\mathfrak{M}_x(a_0 \,|\, \mu) \equiv \tau_x\, \mathfrak{M}(a_0 \,|\, \mu) = \tau_x\, \iota\, \mathfrak{n}(a_0 \,|\, \mu) = (x - a_0) \,|\, \mathfrak{n}\mu \tag{8.11}$$

depending on the orientation the mode is read. Thus, the inversion by x combine, first, an inversion of the directed chord and, afterwards, a translation by an interval x, leading to a new root $x - a_0$.

If the directed chords $a_0 \,|\, \mu$ and $(x - a_0) \,|\, \mathfrak{n}\mu$ are added by intervals, their sum accounts x in each interval. Then, their mean is the same value $\frac{x}{2}$ for each interval, so that their notes are symmetric about $\frac{x}{2}$.

For instance, in this way we may write two directed chords $\tau_x(a_0 \,|\, \mu)$ and $\mathfrak{M}_x(a_0 \,|\, \mu)$ that are symmetric about x.

As a consequence of Eq. 6.7, these operations are not commutative. We notate $\mathfrak{I}_0 = \mathfrak{I}$ and $\mathfrak{M}_0 = \mathfrak{M}$.

Since inversions by an interval are composition of translations and inversions, they are also well-defined operations on chords.

8.4 PROPERTIES

Operations on the root and operations on the mode are mutually independent and commutative. On the other hand, as seen in Eq. 6.7, translations and inversions on the root do not commute. In addition to the property

$$\tau_x \tau_y = \tau_{x+y}$$

it is easy to borne out the following properties involving translations τ_x and inversions $\mathfrak{I}_y$ (or alternatively $\mathfrak{M}_y$):

$$\mathfrak{I}_x \mathfrak{I}_y = \tau_{x-y} \qquad \mathfrak{M}_x \mathfrak{M}_y = \tau_{x-y} \tag{8.12}$$

$$\tau_x \mathfrak{I}_y = \mathfrak{I}_{x+y} = \tau_y \mathfrak{I}_x \qquad \tau_x \mathfrak{M}_y = \mathfrak{M}_{x+y} = \tau_y \mathfrak{M}_x \tag{8.13}$$

$$\mathfrak{I}_x \tau_y = \mathfrak{I}_{x-y} = \tau_{-y} \mathfrak{I}_x \qquad \mathfrak{M}_x \tau_y = \mathfrak{M}_{y-x} = \tau_{-y} \mathfrak{M}_x \tag{8.14}$$

From Eq. 8.12 we see that $\mathfrak{I}_x^2 = \mathfrak{E}$, the identity element.

Since these are operations on chords, we should write the identity element as $\mathfrak{E} = \epsilon\mathfrak{e}$ or $\mathfrak{E} = \tau_0\mathfrak{e}$. Similarly, a translation τ_x on the root of a chord is equivalent to the operation $\tau_x\mathfrak{e}$ on the whole chord, although, for the sake of simplicity, we shall not write the identity element on the mode.

We summarize in Table 8.1 the operations defined for directed chords, acting either on the root, on the mode, or on both, and we point out those well-defined for chords and those that leave the chord invariant. Thus, the non-invariant operations on chords are translations and inversions, which can be characterized in the following form.

Let us remember that, in a n-TET system, the translations and inversions are made by intervals $x \in \mathbb{Z}_n$, so that $\tau_x = \tau_1^x$. The translations, which are generated by τ_1 and satisfy $\tau_1^n = \tau_0$, are isomorphic to $\mathbb{Z}_n$. On the other hand, an inversion by x is obtained as $\mathfrak{I}_x = \tau_x \mathfrak{I}_0$. Since $\mathfrak{I}_0^2 = \tau_0$, then $\mathfrak{I}_0$ generates a group isomorphic to $\mathbb{Z}_2$. Therefore, the group of operations on a chord is generated by τ_1 and $\mathfrak{I}_0$. It is a semidirect product $\mathbb{Z}_n \rtimes \mathbb{Z}_2$ with $2n$ elements $\{\tau_0, \tau_1, \tau_1^2, \ldots, \tau_1^{n-1}, \mathfrak{I}_0, \tau_1\mathfrak{I}_0, \tau_1^2\mathfrak{I}_0, \ldots, \tau_1^{n-1}\mathfrak{I}_0\}$ satisfying $\mathfrak{I}_0\, \tau_1\, \mathfrak{I}_0 = \tau_1{}^{-1}$. Similarly to the operations on the root of a directed chord, the operations on a chord are isomorphic to the dihedral group D_n.

8.5 ROTATIONS

Like inversions, a directed chord rotation, as defined in Eq. 2.22, is a combination of a translation on the root and a shift on the full mode.

Table 8.1 The solid dots mark where the operations are defined and the circles indicate that the chord remains invariant.

	Root	Mode	Chord
ι	•		
τ_x	•		•
$\mathfrak{n}$		•	
$\mathfrak{p}$		•	
$\mathfrak{s}$		•	
$\mathfrak{t}_i$		•	
$\mathfrak{r}$		•	○
$\mathfrak{I}$	•	•	•
$\mathfrak{M}$	•	•	•
$\mathfrak{I}_x$	•	•	•
$\mathfrak{M}_x$	•	•	•
$\mathfrak{D}_i$	•	•	•
$\mathfrak{R}$	•	•	○

A one-step prograde rotation applied to the directed chord $a = a_0 \,|\, \mu \in K(n, k)$ can be expressed depending on the mode intervals in either of the following forms,

$$\mathfrak{R}a = a^1 = \tau_{A_0} a_0 \,|\, \mathfrak{s}\mu = \tau[\mu]_0 \, a_0 \,|\, \mathfrak{s}\mu \tag{8.15}$$

Notice that, for the translation by a mode interval, we are using the notation introduced in Eq. 6.31. Similarly, an i-step prograde rotation applied to the same directed chord, according to Eq. 7.8 and by taking into account Eq. 7.35, can be written as

$$\mathfrak{R}^i a = a^i = \tau_{A_0+\cdots+A_{i-1}} a_0 \,|\, \mathfrak{s}^i \mu = \tau[\mu]_0^i \, a_0 \,|\, \mathfrak{s}^i \mu \tag{8.16}$$

Successive rotations can be notated as follows. The one-step rotation is written as

$$\mathfrak{R}a^i = a^{i+1} = \tau_{A_i} a_i \,|\, \mathfrak{s}\mu^i = \tau[\mu^i]_0 \, a_i \,|\, \mathfrak{s}\mu^i \tag{8.17}$$

and, in general, the consecutive i- and j-step rotations are expressed as

$$\mathfrak{R}^j a^i = a^{i+j} = \tau_{A_i+\cdots+A_{i+j-1}} a_i \,|\, \mathfrak{s}^j \mu^i = \tau[\mu^i]_0^j \, a_i \,|\, \mathfrak{s}^j \mu^i \tag{8.18}$$

Thus, rotations are the result of applying both independent operations on root and mode.

Rotations on directed k-chords clearly satisfy

$$\mathfrak{R}^i a^j = \mathfrak{R}^j a^i = \mathfrak{R}^{i+j} a^0; \quad \mathfrak{R}^k = \mathfrak{R}^0$$

and they have the structure of a cyclic group of order k. Hence, for the rotation index we also assume $\mathfrak{R}^i \equiv \mathfrak{R}^{i \bmod k}$.

Rotations can be defined backwards, as retrograde rotations,

$$\mathfrak{R}^{-1} a^i = a^{i-1} = \tau_{-A_{i-1}} a_i \,|\, \mathfrak{s}^{-1}\mu^i = \tau[\mathfrak{n}\mu^i]_{-1}\, a_i \,|\, \mathfrak{s}^{-1}\mu^i \tag{8.19}$$

Then, an arbitrary anticlockwise rotation is evaluated, by taking into account Eq. 7.39, as

$$\mathfrak{R}^{-j} a^i = a^{i-j} = a_{i-j} \,|\, \mathfrak{s}^{-j}\mu^i = \tau[\mathfrak{n}\mu^i]_{-1}^{-j}\, a_i \,|\, \mathfrak{s}^{-j}\mu^i \tag{8.20}$$

If we take into account Eqs. 6.38 (remember that $\mathfrak{p}\mu = \overline{\mu}$) and 7.4, we may write

$$\tau[\mathfrak{n}\mu^i]_{-1}^{-j} = \tau[\mathfrak{p}\mathfrak{n}\mu^i]_0^j = \tau[\mathfrak{r}\mu^i]_0^j \tag{8.21}$$

allowing us to write the translation in Eq. 8.20 in terms of the prograde intervals of the retrograde mode as

$$\mathfrak{R}^{-j} a^i = a^{i-j} = \tau[\mathfrak{r}\mu^i]_0^j\, a_i \,|\, \mathfrak{s}^{-j}\mu^i \tag{8.22}$$

which is a formula similar to Eq. 8.18, with the only difference that retrograde rotations take translations in the root by the retrograde mode.

The following example shows how to deal with these operators. Let us see that a rotation of a directed chord defined from a negative mode corresponds to a retrograde rotation of the same directed chord defined from the corresponding positive inversion of the mode.

$\mathfrak{R}(a_i \,|\, \mathfrak{n}\mu^i)$ and $\mathfrak{R}^{-1}(a_i \,|\, \mathfrak{p}\mu^i)$, $i \in \mathbb{Z}_k$, provide the same directed chord, although the former expresses the directed chord oriented anticlockwise and the latter, clockwise. Thus, by taking into account Eq. 7.18, prograde rotations, applied to directed chords generated by an inverted mode in the negative form, are calculated as

$$\mathfrak{R}(a^i \,|\, \mathfrak{n}\mu^i) = \tau[\mathfrak{n}\mu^i]_0\, a_i \,|\, \mathfrak{s}\mathfrak{n}\mu^i = (a_i - A_i) \,|\, \mathfrak{n}\mu^{i+1} \tag{8.23}$$

Retrograde rotations, applied to chords generated by an inverted mode in the positive form, are calculated as

$$\mathfrak{R}^{-1}(a_i \,|\, \mathfrak{p}\mu^i) = \tau[\mathfrak{r}\mu^i]_{-1}\, a_i \,|\, \mathfrak{s}^{-1}\mathfrak{p}\mu^i = (a_i - A_i) \,|\, \mathfrak{p}\mu^{i+1} \tag{8.24}$$

which provides the same chord than Eq. 8.23, although read clockwise.

By definition, the chords are invariant under rotations, i.e.,

$$\{a\} = \{\mathfrak{R}^i a^0\}, \ \forall i \in \mathbb{Z}_k \tag{8.25}$$

rotations will induce a constraint between translations and shifts, which is developed in the next section. Furthermore, according to Eq. 7.3, which expresses an inverted chord in two equivalent ways, the relationship $\{\mathfrak{R}^i(a \,|\, \mathfrak{n}\mu)\} = \{\mathfrak{R}^j(a \,|\, \mathfrak{p}\mu)\}$ is satisfied for any couple of indices $i, j \in \mathbb{Z}_k$.

8.6 DRIFTS ALONG EDGES

We now remark a couple of properties involving rotations and the transposition operator. The results obtained in Section 7.5 can be expressed by using a more appropriate notation by defining the *drift* operator, transforming directed chords along single edges of the chord network,

$$\mathfrak{D}_i \equiv \mathfrak{R}^{-i}\mathfrak{t}\mathfrak{R}^i; \quad i \in \mathbb{Z}_k, \ k > 2 \tag{8.26}$$

When acting on the directed chord $a_0 \,|\, \mu = a_0 \,|\, [A_0, \ldots, A_{k-1}]$, bearing in mind Eq. 7.21, we get

$$\mathfrak{D}_i\,(a_0 \,|\, \mu) = \mathfrak{R}^{-i}(a_i \,|\, \mathfrak{t}\mathfrak{s}^i\mu) = a_0 \,|\, \mathfrak{t}_i\mu; \quad i = 0, \ldots, k-2 \tag{8.27}$$

$$\mathfrak{D}_{k-1}\,(a_0 \,|\, \mu) = \mathfrak{R}^{-(k-1)}(a_{k-1} \,|\, \mathfrak{t}\mathfrak{s}^{k-1}\mu) = a_0 + A_0 - A_{k-1} \,|\, \mathfrak{t}_{k-1}\mu \tag{8.28}$$

As a direct consequence of Section 7.5 and of the fact that the transpositions $\{\mathfrak{t}_0, \ldots, \mathfrak{t}_{k-2}\}$ generate the symmetric group $\mathcal{S}_k$, we may affirm,

- ▷ For $i = 0, \ldots, k-2$, the operator $\mathfrak{D}_i$ transforms a directed chord $a_0 \,|\, \mu$ sideways along one edge in the same cell a_0, while for $i = k-1$ the directed chord is sent forward along one edge toward the non-congruent cell $a_0 + A_0 - A_{k-1}$.

In addition,

- ▷ For $j = 2, \ldots, k$, given a subset of $j-1$ consecutive numbers $\{i, i+1, \ldots, i+j-2\}$ in $\mathbb{Z}_k$ and a permutation $\sigma \in \mathcal{S}_{j-1}$, then

$$(\mathfrak{D}_{\sigma(i+j-2)} \cdots \mathfrak{D}_{\sigma(i+1)}\,\mathfrak{D}_{\sigma(i)})^j = \mathfrak{E} \tag{8.29}$$

To proof it, let $\mathfrak{S} = \mathfrak{D}_{\sigma(i+j-2)} \cdots \mathfrak{D}_{\sigma(i+1)}\, \mathfrak{D}_{\sigma(i)}$. When $\mathfrak{S}^j$ is applied to a directed chord $a_0 \,|\, [A_0, \ldots, A_{k-1}]$, with regard to the mode part $\mathfrak{c} = \mathfrak{t}_{\sigma(i+j-2)} \cdots \mathfrak{t}_{\sigma(i+1)}\, \mathfrak{t}_{\sigma(i)}$, Proposition 7.24 guarantees the identity result. With regard to the root, if $\mathfrak{S}$ does not contain $\mathfrak{D}_{k-1}$, the root is also maintained. Otherwise, if $\mathfrak{S}$ contains $\mathfrak{D}_{k-1}$, since $\mathfrak{c}$ is a cycle of length j involving the mode intervals $A_i, A_{i+1}, \ldots, A_{i+j-1}$, the first and the last mode intervals are involved in each iteration and, in particular, from the first application of the operator $\mathfrak{D}_{k-1}$ onward, the root is modified according to Eq. 8.28. Each time $\mathfrak{D}_{k-1}$ is applied, the first mode interval runs a cycle $A_{\rho(0)}, A_{\rho(1)}, \ldots, A_{\rho(j-1)}$ of length j, with $\rho \in \mathcal{S}_j$. Similarly, the last mode interval runs a shifted cycle $A_{\rho(m)}, A_{\rho(m+1)}, \ldots, A_{\rho(m+j-1)}$ for certain $m \neq 0$ in $\mathbb{Z}_k$. Therefore, after j applications of $\mathfrak{D}_{k-1}$, the root a_0 is transformed according to a series of translations $\tau[A_{\rho(0)} - A_{\rho(m)}] \cdots \tau[A_{\rho(j-1)} - A_{\rho(m+j-1)}] = 0$, since all the mode intervals involved in one cycle appear once as positive and once as negative.

Therefore,

▷ For all $i \in \mathbb{Z}_k$, the drift operators satisfy

$$\begin{aligned} &\text{(i)} && \mathfrak{D}_i^2 = \mathfrak{E} \\ &\text{(ii)} && (\mathfrak{D}_j\mathfrak{D}_i)^2 = \mathfrak{E} \Leftrightarrow |i-j| \neq 1 \\ &\text{(iii)} && (\mathfrak{D}_{i+1}\mathfrak{D}_i)^3 = \mathfrak{E} \end{aligned} \tag{8.30}$$

Condition (i) is obvious; (ii) is a consequence of (i), since the transpositions of the mode intervals are disjoint; and (iii) is a consequence of the above result.

8.7 SIMPLE CIRCUITS

Therefore, the elemental polygons composing one cell of k-chords are squares (only if $k > 3$) and hexagons, associated with the simple circuits (closed paths where no vertex is repeated) of the constraints (ii) and (iii) of Eq. 8.30, for indices in $\{0, \ldots, k-2\}$. They are represented in the graphs of Figure 8.1 as wall tiles or bricks of square and hexagonal lattices. Hence, we see that the combination of hexagons and squares that appeared in the cell of tetrachords of Figure 5.5 was not a particular case, but is the pattern for higher dimensions.

Figure 8.1 Wall tiles or bricks composed of square and hexagonal lattices that are contained in the chord cells for $k > 3$.

One node of a chord cell is connected to the neighbor $k-1$ nodes shared with the neighbor cells with the above drifts and the drift $\mathfrak{D}_{k-1}$ connects the node with a non-congruent cell.

In the chord cell there are other simple circuits, such as allowed concatenations of the above ones or paths corresponding to k consecutive shifts of the mode $(\mathfrak{D}_{k-2} \cdots \mathfrak{D}_0)^k$ involving $k(k-1)$ edges. In general, throughout the chord network, we will find the simple circuits described in Eq. 8.29.

Since the Cayley graph is Hamiltonian, in the permutahedron we will always find *Hamiltonian circuits*, i.e., closed paths that go through each vertex of the chord cell exactly once.

For example, in the case of tetrachords, the truncated octahedron with 14 facets admits several Hamiltonian circuits. Figure 8.2 shows the one obtained as

$$\mathfrak{D}_2 \;\; \mathfrak{D}_1\mathfrak{D}_0\mathfrak{D}_1\mathfrak{D}_0\mathfrak{D}_1 \;\; \mathfrak{D}_2\mathfrak{D}_1\mathfrak{D}_2 \;\; \mathfrak{D}_1\mathfrak{D}_0\mathfrak{D}_1\mathfrak{D}_0\mathfrak{D}_1 \;\; \mathfrak{D}_2\mathfrak{D}_1\mathfrak{D}_2 \;\; \mathfrak{D}_1\mathfrak{D}_0\mathfrak{D}_1\mathfrak{D}_0\mathfrak{D}_1 \;\; \mathfrak{D}_2\mathfrak{D}_1$$

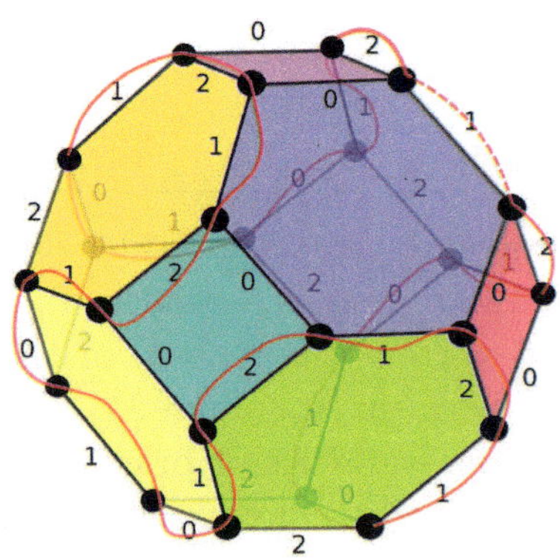

Figure 8.2 Hamiltonian circuit for the cell of tetrachords. Only the indices of $\mathfrak{D}_i$ are notated.

that can start reading beginning in the path marked with a dashed line in either direction.

Notice that the subindices of consecutive drifts are also consecutive, so they do not commute, and the circuits $\mathfrak{c} = (\mathfrak{D}_1\mathfrak{D}_0)^3 = (\mathfrak{D}_0\mathfrak{D}_1)^3 = (\mathfrak{D}_2\mathfrak{D}_1)^3 = (\mathfrak{D}_1\mathfrak{D}_2)^3$ are stopped just before being completed.

8.8 SHORTCUT CIRCUITS

It is also desirable to find the shortest circuits that run along several chord cells allowing changes of tonality, likewise the maximally smooth cycles alternating major and minor triads of the 12-TET scale. Each simple circuit begins and ends in the same chord, so that the total length of the path is equivalent to a number of translations along the chord network satisfying $\sum_i c_i A_i = 0 \bmod n$ for certain values $c_i \in \mathbb{N}$, although the path runs step-by-step along several directed chords of different modes and mode classes (i.e., by directed chords that cannot be achieved only by translations and mode shifts) until becomes closed. It is relevant the case where translations are done always along the same vector interval direction, i.e., when there is a minimum positive integer j, $2 \leq j < n$, satisfying $j\, A_i = 0 \bmod n$. Since, for $j = n$ there is always a simple circuit, we may speak of a *shortcut circuit*. We study the case $j\, A_0 = 0 \bmod n$.

▷ For $i, j \in \mathbb{Z}_k$, the drifts operators satisfy

$$\mathfrak{D}_{i+1} = \mathfrak{R}^{-1}\mathfrak{D}_i\mathfrak{R} \tag{8.31}$$

Hence,

$$\mathfrak{D}_{i+j} = \mathfrak{R}^{-j}\mathfrak{D}_i\mathfrak{R}^j \tag{8.32}$$

For $i = 0, \ldots, k-3$ it is an immediate consequence of Eq. 7.21, since these drifts do not change the root. For $i = k-2$ and $k-1$, the relationship is also satisfied, since

$$\begin{aligned}
&\mathfrak{R}^{-1}\mathfrak{D}_{k-2}\mathfrak{R}(a_0 \,|\, [A_0, \ldots, A_{k-1}]) = \mathfrak{R}^{-1}\mathfrak{D}_{k-2}(a_0 + A_0 \,|\, [A_1, \ldots, A_0]) = \\
&= \mathfrak{R}^{-1}(a + A_0 \,|\, [A_1, \ldots, A_0, A_{k-1}]) = (a + A_0 - A_{k-1} \,|\, [A_{k-1}, A_1, \ldots, A_0]) = \\
&= \mathfrak{D}_{k-1}(a_0 \,|\, [A_0, \ldots, A_{k-1}])
\end{aligned}$$

and

$$
\begin{aligned}
&\mathfrak{R}^{-1}\mathfrak{D}_{k-1}\mathfrak{R}(a_0 \,|\, [A_0, \ldots, A_{k-1}]) = \mathfrak{R}^{-1}\mathfrak{D}_{k-1}(a_0 + A_0 \,|\, [A_1, A_2, \ldots, A_{k-1}, A_0]) = \\
&= \mathfrak{R}^{-1}(a + A_0 + A_1 - A_0 \,|\, [A_0, A_2, \ldots, A_{k-1}, A_1]) = \\
&= (a + A_0 + A_1 - A_0 - A_1 \,|\, [A_1, A_0, A_2, \ldots, A_{k-1}]) = \mathfrak{D}_0(a_0 \,|\, [A_0, \ldots, A_{k-1}])
\end{aligned}
$$

The graph in Eq. 8.33 describes the first translation $\tau[A_0](a \,|\, \mu)$ depending on whether we refer this path to the edges of the adjacent chord cell $a_0 + A_0$ (upper row) or whether it starts running along the edge connecting the cells a_0 and $a_0 + A_0$ (lower row). The translation from the first chord (left column) to the last chord (right column) involves $k-1$ steps.

$$
\begin{array}{ccccccc}
a + A_0 \,|\, \mathfrak{s}\mu & \xrightarrow{\mathfrak{D}_{k-2}} & a + A_0 \,|\, \mathfrak{t}_{k-2}\mathfrak{s}\mu & \xrightarrow{\mathfrak{D}_{k-3}} & \cdots & \xrightarrow{\mathfrak{D}_0} & a + A_0 \,|\, \mu \\
\uparrow \mathfrak{R} & & & \nearrow \tau[A_0] & & & \uparrow \mathfrak{R} \\
a \,|\, \mu & \xrightarrow[\mathfrak{D}_{k-1}]{} & a + A_0 - A_{k-1} \,|\, \mathfrak{t}_{k-1}\mu & \xrightarrow[\mathfrak{D}_{k-2}]{} & \cdots & \xrightarrow[\mathfrak{D}_1]{} & a + A_0 - A_{k-1} \,|\, \mathfrak{s}^{-1}\mu
\end{array}
\tag{8.33}
$$

In the first case, since $\mathfrak{D}_0 \cdots \mathfrak{D}_{k-2} = \mathfrak{s}^{-1}$, $\mathfrak{D}_0 \cdots \mathfrak{D}_{k-2}\mathfrak{R}(a_0 \,|\, \mu) = \mathfrak{s}^{-1}\tau[A_0](a_0 \,|\, \mathfrak{s}\mu) = \tau[A_0](a_0 \,|\, \mu)$, so that $(\mathfrak{D}_0 \cdots \mathfrak{D}_{k-2}\mathfrak{R})^j (a_0 \,|\, \mu) = \tau[j A_0](a_0 \,|\, \mu) = a_0 \,|\, \mu$.

In the second case, by Eq. 8.32, $\mathfrak{R}\mathfrak{D}_1 \cdots \mathfrak{D}_{k-1}(a_0 \,|\, \mu) = \tau[A_0](a_0 \,|\, \mu)$, hence $(\mathfrak{R}\mathfrak{D}_1 \cdots \mathfrak{D}_{k-1})^j (a_0 \,|\, \mu) = \tau[j A_0](a_0 \,|\, \mu) = a_0 \,|\, \mu$. Since chords are invariant under rotations, the path generates a simple circuit along $j\,(k-1)$ chords.

The generalization of the above result to any interval satisfying $j\, A_i = 0 \bmod n$ is straightforward, since it suffices to apply the ith rotation to the directed chord $(a \,|\, \mu)$ at the beginning of the path, and the ith inverse rotation at the end, in order to recover the original mode. Therefore,

▷ If $j\, A_i = 0 \bmod n$, $2 \leq j < n$, there exist shortcut circuits in the chord network running $j\,(k-1)$ chords and edges, so that $(\mathfrak{R}\mathfrak{D}_1 \cdots \mathfrak{D}_{k-1})^j (a_i \,|\, \mu^i) = \tau[j A_i](a_i \,|\, \mu^i) = a_i \,|\, \mu^i$.

Furthermore, by taking into account Eq. 8.32, we may write

$$\cdots(\mathfrak{R}\,\mathfrak{D}_1\cdots\mathfrak{D}_{k-1})(\mathfrak{R}\,\mathfrak{D}_1\cdots\mathfrak{D}_{k-1})(\mathfrak{R}\,\mathfrak{D}_1\cdots\mathfrak{D}_{k-1}) =$$

$$\cdots(\mathfrak{R}\,\mathfrak{R}^2\mathfrak{R}^{-2}\mathfrak{D}_1\mathfrak{R}^2\mathfrak{R}^{-2}\cdots\mathfrak{R}^2\mathfrak{R}^{-2}\mathfrak{D}_{k-1})(\mathfrak{R}\,\mathfrak{R}\mathfrak{R}^{-1}\mathfrak{D}_1\mathfrak{R}\mathfrak{R}^{-1}\cdots\mathfrak{R}\mathfrak{R}^{-1}\mathfrak{D}_{k-1})$$

$$(\mathfrak{R}\,\mathfrak{D}_1\cdots\mathfrak{D}_{k-1}) = \cdots(\mathfrak{D}_3\cdots\mathfrak{D}_0\mathfrak{D}_1)(\mathfrak{D}_2\cdots\mathfrak{D}_{k-1}\mathfrak{D}_0)(\mathfrak{D}_1\cdots\mathfrak{D}_{k-1})$$

Therefore,

▷ The shortcut circuit of the above result can be run without rotations, by consecutive application of drifts with decreasing indices in $\mathbb{Z}_k$ (this is also valid for any simple circuit),

$$\underbrace{\cdots(\mathfrak{D}_3\cdots\mathfrak{D}_0\mathfrak{D}_1)(\mathfrak{D}_2\cdots\mathfrak{D}_{k-1}\mathfrak{D}_0)(\mathfrak{D}_1\cdots\mathfrak{D}_{k-1})}_{j(k-1)\ steps}(a_i \,|\, \mu^i) = a_i \,|\, \mu^i$$

CHAPTER 9

Chord network

9.1 SOME FAMILIES OF CHORDS

For a root $a_0 \in \mathcal{O}_n$ and a full mode $\mu \equiv \mu^0 = [A_0, \ldots, A_{k-1}] \in M(n,k)$, we point out some examples of families of chords.

• $F_1(a_0, \mu)$: chords formed as $\{a_0 \,|\, \mu^i\}$, $i \in \mathbb{Z}_k$, with the same cardinality than μ^S.

The family F_1 contains the chords generated from the root a_0 by applying the shifts of the mode μ. In the case of trichords, $k = 3$, the family $F_1(a_0, \mu)$ is the set of three chords of Figure 4.6, marked with black circles within a hexagonal chord cell, which also correspond to the chords of Eq. 4.7, with the same root and the three mode shifts of μ. The family $F_1(a_0, \overline{\mu})$, with the inverted mode of μ, is the set of three chords marked with white circles within a hexagonal chord cell, corresponding to the chords of Eq. 4.8 generated by the mode class of $\overline{\mu}$. Therefore, for trichords, $F_1(a_0, \mu) \cup F_1(a_0, \overline{\mu})$ is the whole chord cell a_0, containing, in general, the six chords generated by the mode classes of μ and $\overline{\mu}$. However, for $k > 3$, the chord cell $\{a_0 \,|\, \mu\}$ is also generated by other interval permutations of the mode μ, besides its shifts and inversions.

• $F_2(a_0, \mu)$: chords formed as $\{\tau_{A_i}\ a_0 \,|\, \mu\}$ and $\{\tau_{-A_i}\ a_0 \,|\, \mu\}$ with $i \in \mathbb{Z}_k$, of cardinality $2k$ if $2A_i \neq n$.

F_2 is the set of chords generated by the mode μ in the neighbor chord cells of a_0 that are adjacent to its front and back facets, respectively. In the new chord cells, the chords are placed in the same relative position than the chord $\{a_0 \,|\, \mu\}$ with regard to the cell a_0. For example, for $k = 3$, according to Figure 4.6, $F_2(a_0, \mu)$ contains the six black circles around the chord $\{a_0 \,|\, \mu\}$.

• $F_3(a_0, \mu)$: chords formed as $\{a_i \,|\, \mu\}$, $i \in \mathbb{Z}_k$, of cardinality k.

The family F_3 is composed of chords which root is a note of the

DOI: 10.1201/9781003597964-9

chord $\{a_0 \,|\, \mu\}$ and are generated by the same mode. Hence, in each cell of chords centered at a_i, which is the root of a rotation of the directed chord $a_0 \,|\, \mu$, the chords of F_3 are placed in a similar relative position.

• $F_4(a_0, \mu)$: chords formed in the following form. Starting from the directed chord $a = a_0 \,|\, \mu$, $\mu \in M(n,k)$, we decompose every shift of the mode as $\mu^i = \beta^i \cdot \gamma^i$, with $\dim(\gamma^i) = 2$, by forming a new mode $\nu^i = \beta^i \cdot \mathfrak{p}\gamma^i$, where $\mathfrak{p}\gamma^i$ is the inverted mode of γ^i. By this way, the chord $\{a\}$ and each chord $\{b^i\} = \{a_i \,|\, \beta^i \cdot \mathfrak{p}\gamma^i\}$ have at least $k-1$ notes in common, which are those of the subchord $\{a_i \,|\, [\beta^i, *]\}$. Therefore, the chords $\{b^i\}$, k in number, are at one step from $\{a\}$ in the network of chords and share an edge with $\{a\}$.

• $F_5(a_0, \mu)$: chords formed as $\{a_0 \,|\, \nu\}$, with $\nu \in \mathcal{S}(\mu)$. This is the cell of chords centered in a_0.

• $F_6(a_0, \mu)$: chords formed as $\{a_i \,|\, \nu\}$, with $i \in \mathbb{Z}_k$ and $\nu \in \mathcal{S}(\mu)$. Since the roots are the notes of the chord $\{a_0 \,|\, \mu\}$, F_6 is composed of the chords belonging to the conglomerate of cells around the cell a_0.

In the following sections these families are studied in more detail.

9.2 REFERRING A CHORD TO DIFFERENT CELLS

As we know, a directed chord rotation consists of two independent operations, a translation on the root and a shift on the mode, that leave the chord invariant. For a directed chord $x_0 \,|\, \mu$, the condition of invariance of the chord, $\{x_0 \,|\, \mu\} = \{\mathfrak{R}(x_0 \,|\, \mu)\}$, according to Eq. 8.17, implies

$$\{x_0 \,|\, \mu\} = \{\tau[\mu]_0 \, x_0 \,|\, \mathfrak{s}\mu\} \tag{9.1}$$

Then, by applying the opposite translation $\tau^{-1}[\mu]_0 = \tau[-\mu]_0$ to the chords in both sides of the foregoing equation, we get

$$\{\tau^{-1}[\mu]_0 \, x_0 \,|\, \mu\} = \{x_0 \,|\, \mathfrak{s}\mu\} \tag{9.2}$$

Therefore, starting from the chord $\{x_0 \,|\, \mu\}$ we may reach the chord $\{x_0 \,|\, \mathfrak{s}\mu\}$ either from a shift or from a translation. In a similar way as in the geometric representation of Figure 4.6, the translated chord in the new cell is placed at the same relative position than the chord $\{x_0 \,|\, \mu\}$, while the shifted chord remains in the original chord cell x_0, although in a different relative position. Therefore, the same chord is expressed from two roots, which are the center of two contiguous cells sharing the chord of Eq. 9.2.

According to Eqs. 8.15 and 8.25, a chord $\{x_0 \,|\, \nu\}$ of the family $F_5(x_0, \mu)$ (i.e., the chord cell centered in x_0) is invariant by rotations, hence

$$\{x_0 \,|\, \nu\} = \{\mathfrak{R}^j(x_0 \,|\, \nu)\} = \{\tau[\nu]_0^j \, x_0 \,|\, \mathfrak{s}^j \nu\}, j \in \mathbb{Z}_k \tag{9.3}$$

For each chord, the value $j = 0$ provides the chord written as a directed chord with root x_0, while values $j = 1, \ldots, k-1$ provide the same chord from the other roots. Then, each chord $\{x_0 \,|\, \nu\}$ can be expressed according to a directed chord referred to one of the $k-1$ surrounding cells of x_0 centered in $\tau[\nu]_0^j \, x_0$, although from a different shift of the mode.

For the same reason, each relative mirror chord $\{x_0 \,|\, -\nu\}$, which is the opposite chord of $\{x_0 \,|\, \nu\}$ through x_0, can be expressed from the root of its congruent cells centered in $x_j = \tau^{-1}[\nu]_0^j \, x_0$. According to Eq. 7.39, these cells are the successive notes of the relative mirror chord, read anticlockwise. Thus, by generalizing Eq. 9.2, we are led to the following result.

▷ In a chord cell, the chords $\{x_0 \,|\, \mathfrak{s}^j \nu\}$, $j \in \mathbb{Z}_k$, are equivalent to the following translations on the chord $\{x_0 \,|\, \nu\}$,

$$\{x_0 \,|\, \mathfrak{s}^j \nu\} = \{\tau^{-1}[\nu]_0^j \, x_0 \,|\, \nu\} \tag{9.4}$$

The translated chords become referred to directed chords in neighbor cells according to the successive notes of the relative mirror chord $\{x_0 \,|\, \mathfrak{n}\nu\}$, read anticlockwise.

By using dependent translations on the mode components, Eq. 9.4 is expressed as

$$\{\mathfrak{s}^j(x_0 \,|\, \nu)\} = \{\tau^{-1}[\cdot]_0^j \, (x_0 \,|\, \nu)\} \tag{9.5}$$

Similarly, by taking into account Eq. 8.22, the retrograde rotations of a chord satisfy

$$\{x_0 \,|\, \nu\} = \{\mathfrak{R}^{-j}(x_0 \,|\, \nu)\} = \{\tau^{-1}[\mathfrak{p}\nu]_0^j \, x_0 \,|\, \mathfrak{s}^{-j} \nu\} \tag{9.6}$$

By applying $\tau[\mathfrak{p}\nu]_0^j$ in both sides, we get an equivalent result to the previous one.

▷ In a chord cell, the chords $\{x_0 \,|\, \mathfrak{s}^{-j} \nu\}$, $j \in \mathbb{Z}_k$, are equivalent to the following translations on the chord $\{x_0 \,|\, \nu\}$,

$$\{x_0 \,|\, \mathfrak{s}^{-j} \nu\} = \{\tau[\mathfrak{p}\nu]_0^j \, x_0 \,|\, \nu\} = \{\tau[\nu]_{-1}^{-j} \, x_0 \,|\, \nu\} \tag{9.7}$$

Notice that the translations may be rewritten according to Eq. 7.34, so that they are carried out by the same mode ν. Now, the translated chords are referred to directed chords in neighbor cells according the successive notes of the relative inverted chord $\{x_0 \,|\, \mathfrak{p}\mu\}$.

Bearing in mind Eq. 7.34, it can also be written by using dependent translations, as

$$\{\mathfrak{s}^{-j}(x_0 \,|\, \nu)\} = \{\tau[\cdot]_{-1}^{-j}\,(x_0 \,|\, \nu)\} \tag{9.8}$$

Thus, the directed chords of a cell may be referred to different roots depending on whether a shift or a translation is carried out.

For the chords generated by an inverted mode, by taking into account that $-\mathfrak{p} = \mathfrak{n}\mathfrak{p} = \mathfrak{r}$ and $\mathfrak{p}^2 = \mathfrak{e}$, we get some similar results.

▷ The chords $\{x_0 \,|\, \mathfrak{s}^j\, \mathfrak{p}\nu\}$ in a chord cell satisfy

$$\{x_0 \,|\, \mathfrak{s}^j\, \mathfrak{p}\nu\} = \{\tau^{-1}[\mathfrak{p}\nu]_0^j\, x_0 \,|\, \mathfrak{p}\nu\} \tag{9.9}$$

The translated chords are referred to directed chords in neighbor cells according the successive notes, anticlockwise direction, of the chord $\{x_0 \,|\, -\mathfrak{p}\nu\} = \{x_0 \,|\, \mathfrak{r}\nu\}$, obtained from the retrograde mode.

▷ The chords $\{x_0 \,|\, \mathfrak{s}^{-j}\, \mathfrak{p}\nu\}$, $j \in \mathbb{Z}_k$, in a chord cell satisfy

$$\{x_0 \,|\, \mathfrak{s}^{-j}\, \mathfrak{p}\nu\} = \{\tau[\nu]_0^j\, x_0 \,|\, \mathfrak{p}\nu\} = \{\tau[\mathfrak{p}\nu]_{-1}^{-j}\, x_0 \,|\, \mathfrak{p}\nu\} \tag{9.10}$$

Here, the translation may be also rewritten according to Eq. 7.34, so that it is carried out by the same mode $\mathfrak{p}\nu$. The translated chords become referred to directed chords in neighbor chord cells according the successive notes of the chord $\{x_0 \,|\, \nu\}$.

9.3 CO-CYCLES, CO-CELLS, AND CONGRUENT CELLS

For a generic mode $\mu \in M(k, n)$ we write Eq. 9.4 as

$$\{\tau^{-1}[\mu]_0^j\, x_0 \,|\, \mu\} = \{x_0 \,|\, \mu^j\}, \quad j \in \mathbb{Z}_k \tag{9.11}$$

In the cell of chords x_0, the family of chords described by the right-hand side member of Eq. 9.11 is composed of k different chords, which form a *co-cycle* [10]. The co-cycles of the cell x_0 are then generated by the modes of the mode class μ^S applied to the same root, i.e., are the family $F_1(x_0, \mu)$.

Alternatively, on the left-hand side of this equation, each chord of the co-cycle in the cell x_0 is referred to one of the $k-1$ surrounding cell of x_0 with centers in $\tau^{-1}[\mu]_0^j\, x_0$, respectively, in addition to the chord $\{x_0 \,|\, \mu\}$. We call them *co-cells* of the chord $\{x_0 \,|\, \mu\}$, where the chords of the co-cycle are placed *in a similar relative position* as $\{x_0 \,|\, \mu\}$ with regard to the cell x_0.

Similarly, for the inverted mode, Eq. 9.9 written for a mode μ is

$$\{\tau[\mathfrak{r}\mu]_0^j\, x_0 \,|\, \mathfrak{p}\mu\} = \{x_0 \,|\, (\mathfrak{p}\mu)^j\}, \quad j \in \mathbb{Z}_k \tag{9.12}$$

On the left-hand side, the new roots are the successive notes anticlockwise direction of the chord $\{x_0 \,|\, \mathfrak{r}\mu\}$, obtained from the retrograde mode. Then, we get the co-cycle of the family $F_1(x_0, \mathfrak{p}\mu)$. On the left-hand side of the above equation, the chords of the co-cycle in the cell x_0 are referred to the $k-1$ neighbor co-cells of $\{x_0 \,|\, \mathfrak{p}\mu\}$, in addition to the cell x_0. In these cells with centers in $\tau[\mathfrak{r}\mu]_0^j\, x_0$, respectively, these chords occupy the same relative position than $\{x_0 \,|\, \mathfrak{p}\mu\}$ in x_0.

Two chords $\{a\}$ and $\{b\}$ in the cell x_0 are in the same co-cycle if, and only if, for any interval u the translated chords $\{\tau_u\, a\}$ and $\{\tau_u\, b\}$ are in the same co-cycle in the cell $x_0 + u$.

On the other hand, in the cell x_0 a directed chord $x_0 \,|\, \mu$ is placed in a vertex shared by k co-cells. In each one of these cells this chord is generated by other shifts μ^m of the mode[1]. Since the cells sharing the chord are those corresponding to the k rotations of the directed chord, according to Eq. 8.18 with $i = 0$, the chord $\{x_0 \,|\, \mu\}$ in the cell $x_j = \tau[\mu]_0^j\, x_0$ (which is the root of the j-th prograde rotation of the directed chord) is generated by the shift μ^j, for $j = 1, \ldots, k-1$. Then, the shift of the mode is $m = j$ for the cell x_j.

The cell x_0 containing the chord $\{x_0 \,|\, \mu\}$ also contains its relative inverted chord $\{x_0 \,|\, -\mu\}$, as well as their co-cycles. Furthermore, the cell x_0 is contiguous to the cells centered at the notes of all of these chords.

Certainly, for $k > 3$ this result can be generalized to any mode class of $\mathcal{C}(\mu)$ besides those of the modes μ and $\mathfrak{p}\mu$. For any mode class $\nu^S \in \mathcal{C}(\mu)$, the co-cycle $\{x_0 \,|\, \nu^i\}$, $i = 0, \ldots, k-1$, belong to the cell x_0 (as well as their relative inverted chords $\{x_0 \,|\, -\nu^i\}$ generated from the modes of the class $\overline{\nu}^S$, also in $\mathcal{C}(\mu)$). If the Tonnetz is non-degenerate, the co-cycles of a cell have different sets of chords.

[1] In a non-degenerate Tonnetz, these shifts are always different, otherwise, for $p \neq q$, the shifts μ^p and μ^q may coincide.

By collecting all the modes of the mode classes in $\mathcal{C}(\mu)$ we get the class of the symmetric modes $\mathcal{S}(\mu)$. Then, the congruent cells of the cell x_0 are those of the notes composing the chords $\{x_0 \,|\, \nu\}$ with $\nu \in \mathcal{S}(\mu)$. According to Section 5.7, this makes a number of $2^k - 2$ cells, that are adjacent to the facets of the cell x_0.

9.4 DEPENDENT OPERATIONS ON CHORDS

According to Table 8.1, transformations on modes of directed chords are operations not defined on chords. We have seen some examples, like shifts, where the operations do not transform in the same way all the directed chords of a chord class, and they had to be applied within one chord cell, for a fixed root. However, Eqs. 9.5 and 9.8 show that a mode shift can be computed from a dependent translation by mode intervals of a directed chord, which is an operation that, in this form, allows to refer the directed chord to another cell. For example, in the case described by Eq. 7.11, the mode may be shifted according to Eq. 9.5 as follows[2]

$$\{a_0 \,|\, \mu^0\} = \{a_1 \,|\, \mu^1\} \Rightarrow \{\tau^{-1}[\mu^0]_0 \,(a_0 \,|\, \mu^0)\} = \{\tau^{-1}[\mu^0]_0 \,(a_1 \,|\, \mu^1)\} \tag{9.13}$$

In the second member of the latter equation, we get the shift of $\{a_0 \,|\, \mu^0\}$ in the cell a_0:

$$\{\tau^{-1}[\mu^0]_0 \,(a_1 \,|\, \mu^1)\} = \{a_1 - A_0 \,|\, \mu^1\} = \{a_0 \,|\, \mu^1)\}$$

while, in the first member, the shift is expressed as a translation to the cell $a_0 - A_0$, such as

$$\{\tau^{-1}[\mu^0]_0 \,(a_0 \,|\, \mu^0)\} = \{a_0 - A_0 \,|\, \mu^0\}$$

It corresponds to a chord placed in the same relative cell position than $\{a_0 \,|\, \mu^0\}$.

Therefore, we may use the alternative operations described in Eqs. 9.5 and 9.8 to compute the chords obtained as shifts of a directed chord rotation $a_i \,|\, \mu^i$.

[2]If we apply dependent translations to two rotations of the same chord, we get

$$\{\tau^{-1}[\cdot]_0 \,(a_0 \,|\, \mu^0)\} \neq \{\tau^{-1}[\cdot]_0 \,(a_1 \,|\, \mu^1)\}$$

since the translations are made by different mode intervals on both sides.

▷ For any rotation of the directed chord $a_i \,|\, \mu^i$, $i \in \mathbb{Z}_k$, the chords corresponding to the shifted directed chords are obtained from the following translations on the chord $\{a\}$,

$$\begin{aligned} \{\mathfrak{s}^j(a_i \,|\, \mu^i)\} &= \{\tau^{-1}[\cdot]_0^j \; a_i \,|\, \mu^i\} = \tau^{-1}[\mu^i]_0^j \; \{a\} \\ \{\mathfrak{s}^{-j}(a_i \,|\, \mu^i)\} &= \{\tau[\cdot]_{-1}^{-j} \; a_i \,|\, \mu^i\} = \tau[\mu^i]_{-1}^{-j} \; \{a\} \end{aligned} \tag{9.14}$$

We have a similar situation in Eq. 6.10, for inversions on the root of a directed chord. The inversion of the root can be obtained as a translation also depending on the root, that can be expressed as $\{\iota x \,|\, \mu\} = \tau_x^{-2}\{x \,|\, \mu\}$. Thus,

▷ The chord obtained as a root inversion of the ith rotation $a_i \,|\, \mu^i$ is the result of the following translation on the chord $\{a\}$,

$$\{\iota(a_i \,|\, \mu^i)\} = \{\tau^{-2}[\cdot] \; a_i \,|\, \mu^i\} = \tau^{-2}[a_i] \; \{a\} \tag{9.15}$$

Similarly, by taking into account Eq. 8.3 and the above equation,

▷ The chord obtained as the relative mirror chord of $a_i \,|\, \mu^i$ is computed from the following composition on the chord $\{a\}$,

$$\{\mathfrak{n}(a_i \,|\, \mu^i)\} = \{\mathfrak{M} \; \tau^{-2}[\cdot] \; a_i \,|\, \mu^i\} = \mathfrak{M} \; \tau^{-2}[a_i] \; \{a\} \tag{9.16}$$

which is equivalent to the relative inverted chord

$$\{\mathfrak{p}(a_i \,|\, \mu^i)\} = \{\mathfrak{I} \; \tau^{-2}[\cdot] \; a_i \,|\, \mu^i\} = \mathfrak{I} \; \tau^{-2}[a_i] \; \{a\} \tag{9.17}$$

In this way, operations acting only on the root or the mode of a directed chord can also be applied to chords.

Let us see an example to illustrate the use of foregoing relationships. We shall find the chord symmetric to $\{a\} = \{a_i \,|\, \mu^i\}$ about a note y (not necessarily in the chord). Bearing in mind that the relative mirror chord by x of the directed chord $a_i \,|\, \mu^i$ is $a_i + x \,|\, \mathfrak{n}\mu^i$, which are symmetric about $a + \frac{x}{2}$, we should calculate the relative mirror chord by $x = 2y - 2a$, which is the chord

$$\{2y - a_i \,|\, \mathfrak{n}\mu^i\} = \{\tau[2y]\mathfrak{M}(a_i \,|\, \mu^i)\} = \{\mathfrak{M}\tau^{-2}[y](a_i \,|\, \mu^i)\}$$

Since these operations are well defined for chords, they can be evaluated from any rotation of the directed chord a, so that

$$\{2y - a_i \,|\, \mathfrak{n}\mu^i\} = \mathfrak{M} \; \tau^{-2}[y] \; \{a\}$$

In particular, for the chord symmetric about the note a_j composing the chord $\{a\}$, we have

$$\mathfrak{M}\,\tau^{-2}[a_j]\,\{a\} = \mathfrak{M}\,\tau^{-2}[a_j]\,\{a_j\,|\,\mu^j\} = \{a_j\,|\,\mathfrak{n}\mu^j\}$$

so that we get back to Eq. 9.17.

9.5 SINGLE TRANSLATIONS

We study in detail the transformations on the Tonnetz between chords with the same mode class, i.e., those which mode is a shift of a particular mode. As we have seen, these transformations can be interpreted as consecutive single translations of chords. To this purpose, it is necessary to assume that the Tonnetz is non-degenerate, i.e., that the interval vectors have different length.

According to the preceding section, the chord $\{x_0\,|\,\mu^0\}$ in the cell x_0 belongs to the co-cycle formed by k chords (we assume $k \geq 3$) in the form $\{x_0\,|\,\mu^i\}$, $i \in \mathbb{Z}_k$. In the cell x_0, these chords can be expressed from directed chords with root x_0 and shifts of the mode class μ^S and they may also be expressed as translations to the co-cells by consecutive intervals of a mode of the class. Among these translations, there are only two translations depending on a *single* mode interval, the first and the last, yielding a chord of the co-cycle in the same cell. According to Eq. 9.14, for the directed chord $x_0\,|\,\mu^i$, these translations are

$$\begin{aligned} \tau[-\mu^i]_0\,\{x_0\,|\,\mu^i\} &= \{x_0\,|\,\mu^{i+1}\} \\ \tau[\mu^i]_{-1}\,\{x_0\,|\,\mu^i\} &= \{x_0\,|\,\mu^{i-1}\} \end{aligned} \tag{9.18}$$

Other translations by a single interval lead to a chord in another cell.

Thus, if the translation is carried out by an arbitrary mode interval $[-\mu^i]_j = [-\mu^{i+j}]_0 = -A_{i+j}$, for $0 < j \leq k-1$, by expressing the chord from the j-th prograde rotation of $x_0\,|\,\mu^i$, i.e., $\{\mathfrak{R}^j(x_0\,|\,\mu^i)\} = \{x_j\,|\,\mu^{i+j}\}$, the first expression of Eq. 9.14 allows us to use a translation as a shift, so that

$$\begin{aligned} &\tau[-\mu^{i+j}]_0\{x_0\,|\,\mu^i\} = \tau[-\mu^{i+j}]_0\{\mathfrak{R}^j(x_0\,|\,\mu^i)\} = \\ &= \tau[-\mu^{i+j}]_0\{x_j\,|\,\mu^{i+j}\} = \{x_j\,|\,\mu^{i+j+1}\} \end{aligned} \tag{9.19}$$

Therefore, the chord $\{x_j\,|\,\mu^{i+j+1}\}$ in the cell x_j is not a rotation of any chord in the cell x_0, although in the cell x_j there is the chord $\{x_j\,|\,\mu^{i+j}\}$ which is a rotation of $x_0\,|\,\mu^i$.

Then, these $k-1$ translations

$$\tau[-\mu^{i+j}]_0 \{x_0 \,|\, \mu^i\} = \{x_j \,|\, \mu^{i+j+1}\}, \quad 0 < j \leq k-1 \tag{9.20}$$

lead to a chord which is not in the co-cycle of $\{x_0 \,|\, \mu^i\}$ and is placed in the neighbor cell of x_0

$$x_j = x_0 + A_0 + \cdots + A_{j-1}$$

Similarly, for the second relationship of Eq. 9.18, if the translation is carried out by the mode interval $[\mu^i]_{-j} = [\mu^{i-j+1}]_{-1} = A_{i-j}$, for $1 < j \leq k$, by expressing the chord from the $(j-1)$-th retrograde rotation of $x_0 \,|\, \mu^i$, i.e., $\{\mathfrak{R}^{-(j-1)}(x_0 \,|\, \mu^i)\} = \{x_{k-j+1} \,|\, \mu^{i-j+1}\}$, and taking into account Eq. 9.14, we get

$$\begin{aligned}&\tau[\mu^{i-j+1}]_{-1}\{\mathfrak{R}^{-(j-1)}(x_0 \,|\, \mu^i)\} = \\ &= \tau[\mu^{i-j+1}]_{-1}\{x_{k-j+1} \,|\, \mu^{i-j+1}\} = \{x_{k-j+1} \,|\, \mu^{i-j}\}\end{aligned} \tag{9.21}$$

Therefore, the chord $\{x_{k-j+1} \,|\, \mu^{i-j}\}$ in the cell x_{k-j+1} cannot be expressed as a rotation of a directed chord in the cell x_0, although in the same cell there is the chord $\{x_{k-j+1} \,|\, \mu^{i-j+1}\}$ which is a rotation of $x_0 \,|\, \mu^i$.

Then, these $k-1$ translations

$$\tau[\mu^{i-j+1}]_{-1} \{x_0 \,|\, \mu^i\} = \{x_{k-j+1} \,|\, \mu^{i-j}\}, \quad 1 < j \leq k \tag{9.22}$$

lead to a chord that is not in the co-cycle of $\{x_0 \,|\, \mu^i\}$ and is placed in the neighbor cell of x_0

$$x_{k-j+1} = x_0 - A_{k-1} - \cdots - A_{k-j+1}$$

The Eqs. 9.18, 9.20, and 9.22, written in terms of the intervals of the mode $\mu = [A_0, \ldots, A_{k-1}]$, can be summarized as follows.

▷ The chords resulting from applying a single translation by a mode interval to the chord $\{x_0 \,|\, \mu\}$ are,

$$\tau[-A_0]\,\{x_0 \,|\, \mu\} = \{x_0 \,|\, \mu^1\}; \; \tau[A_{k-1}]\,\{x_0 \,|\, \mu\} = \{x_0 \,|\, \mu^{-1}\} \tag{9.23}$$

and

$$\begin{aligned}\tau[-A_l]\,\{x_0 \,|\, \mu\} &= \{x_l \,|\, \mu^{l+1}\}; \quad 1 \leq l \leq k-1 \\ \tau[A_l]\,\{x_0 \,|\, \mu\} &= \{x_{l+1} \,|\, \mu^l\}; \quad 0 \leq l \leq k-2\end{aligned} \tag{9.24}$$

The two translations of Eq. 9.23 remain in the same cell. The ones of Eq. 9.24 cannot be rooted in the cell x_0 and describe the chords from congruent cells, although in the left-hand side the chords are rooted in a note not belonging to the original chord and on the right-hand side they have a root that is a note of the original chord.

▷ Single translations by a mode interval of the chord $\{x_0 \,|\, \mu\}$ are composed of the chords $\{x_l \,|\, \mu^{l+1}\}$ with $0 \leq l \leq k-1$, and the chords $\{x_{l+1} \,|\, \mu^l\}$ with $0 \leq l \leq k-1$.

This corresponds to the family $F_2(x_0, \mu)$.

To complete the study of single translations of a chord $\{x_0 \,|\, \mu\}$, we will investigate when two single translations lead to chords sharing a common chord cell, either the same cell x_0 or a different one, and when they drive to different cells, that is, that they cannot share a common root. This will be done by comparing the pairs of possible positive and negative translations, i.e., by $\tau[-A_p]$ and $\tau[A_q]$, by $\tau[A_p]$ and $\tau[A_q]$, and by $\tau[-A_p]$ and $\tau[-A_q]$.

9.6 TRANSLATIONS TOWARD ONE CELL

According to Eq. 9.24, a translation by a single mode interval of a chord $\{x_0 \,|\, \mu\}$ in the cell x_0 leads to a cell different from x_0 in $2k-2$ out of $2k$ cases. Nevertheless, different single translations of the chord $\{x_0 \,|\, \mu\}$ may lead to a common cell, that is, to chords belonging to one co-cycle.

Let us see in which cases the families of translated chords

$$\begin{aligned} \tau[-A_p]\;\{x_0 \,|\, \mu\} &= \{x_p \,|\, \mu^{p+1}\}; \quad p \in \mathbb{Z}_k \\ \tau[A_q]\;\{x_0 \,|\, \mu\} &= \{x_{q+1} \,|\, \mu^{q}\}; \quad q \in \mathbb{Z}_k \end{aligned} \tag{9.25}$$

produce two chords that can be referred to the same root.

We already know that for $p = 0$ and $q = k-1$, the translations leave the chords in the cell x_0 according to Eq. 9.23. It is also obvious that for $q = p - 1 \bmod k$, the translations by $\tau[-A_p]$ and $\tau[A_{p-1}]$ lead to the cell x_p. This includes the case $p = 0$. Therefore,

▷ The chord translations

$$\begin{aligned} \{\tau[-A_p]\;x_0 \,|\, \mu\} &= \{x_p \,|\, \mu^{p+1}\}; \quad p \in \mathbb{Z}_k \\ \{\tau[A_{p-1}]\;x_0 \,|\, \mu\} &= \{x_p \,|\, \mu^{p-1}\}; \quad p \in \mathbb{Z}_k \end{aligned} \tag{9.26}$$

lead to a common neighbor cell.

Nevertheless, in the general case, we will ask for values p and q so that both chords of Eq. 9.25 will have rotations

$$\{\mathfrak{R}^i\, \tau[-A_p]\, x_0 \,|\, \mu\}\ ; \quad \{\mathfrak{R}^{-j}\, \tau[A_q]\, x_0 \,|\, \mu\} \tag{9.27}$$

with a common root for some values i, j, such that $0 \leq i < k$ and $0 \leq j < k$. Then, by Eqs. 8.18 and 8.20,

$$\begin{aligned} \{\mathfrak{R}^i\, \tau[-A_p]\, x_0 \,|\, \mu\} &= \{\tau[\mu]_0^i\, \tau[-A_p]\, x_0 \,|\, \mathfrak{s}^i\, \mu\} \\ \{\mathfrak{R}^{-j}\, \tau[A_q]\, x_0 \,|\, \mu\} &= \{\tau[-\mu]_{-1}^{-j}\, \tau[A_q]\, x_0 \,|\, \mathfrak{s}^{-j}\, \mu\} \end{aligned} \tag{9.28}$$

Thus, by relating the part corresponding to the root of both equations, after reordering terms, we write[3]

$$A_{k-j} + A_{k-j+1} + \cdots + A_{k-1} + A_0 + \cdots + A_{i-2} + A_{i-1} = A_p + A_q \mod n \tag{9.29}$$

The intervals on both sides are positive and on the left there are at most $2k-2$ terms[4]. Then, the values i, j allowing the above relationship are as follows.

The simplest case corresponds to $i = 1$ and $j = 1$, so that either $p = 0$, $q = k-1$ or $p = k-1$, $q = 0$.

Other cases are those involving all the intervals of one full k-mode, i.e., any shift of μ, since $A_0 + \cdots + A_{k-1} = 0 \mod n$. Then, Eq. 9.29 is satisfied if, and only if,

$$A_{i-2} + \underbrace{A_{i-1} + \cdots + A_{k-1} + A_0 + \cdots + A_{i-2}}_{0 \bmod n} + A_{i-1} = A_p + A_q \mod n \tag{9.30}$$

[3] It is worth noticing that this is a way to test whether two chords $\{x \,|\, \mu\}$ and $\{x' \,|\, \mu'\}$ belong to the same co-cycle. Since it is not sure that both chords may be expressed as directed chords with the same root (in which case we should only to compare whether both modes belong to the same class), we write both chords as generated by the same mode $\mu = [A_0, \ldots, A_{k-1}]$, e.g., $\{x' \,|\, \mu'\} = \{y \,|\, \mu\}$. Then, both chords are in the same co-cycle if, and only if, there exist two indices i, j so that the interval difference among the roots x and y can be expressed as a series $A_{k-j} + A_{k-j+1} + \cdots + A_{k-1} + A_0 + \cdots + A_{i-2} + A_{i-1}$, i.e., as a fraction of any shift of the mode μ.

[4] Let us remember that this equality has a geometric meaning. According to Eq. 5.4, the series of intervals $A_0, \ldots, A_{i-1}$ ($i < k$) are associated with independent directions. Similarly, the series $A_{k-j}, \ldots, A_{k-1}$ ($j < k$) are independent vector intervals. A number of k consecutive terms are dependent and form a cycle, hence they can be suppressed. In that case, the remaining terms on the left-hand side are at most $k-2$ in number and are independent. In addition, the intervals A_p and A_q must be included among those terms.

so that the underbraced terms do not contribute to the total sum. Hence, comparing to Eq. 9.29, we get $k-j=i-2 \mod k$. Therefore, since the indices of the mode intervals are different, one of the following cases is hold:

(a) $p=i-1; \quad q=k-j \Longrightarrow q=p-1; \quad i=p+1; \quad -j=p-1$

(b) $p=k-j; \quad q=i-1 \Longrightarrow q=p+1; \quad i=p+2; \quad -j=p$

All these equalities are mod k.

Case (a) is similar to Eq. 9.26. Case (b) provides the family of couples of chords

$$\begin{aligned} \{\tau[-A_p]\, x_0 \,|\, \mu\} &= \{x_p \,|\, \mu^{p+1}\}; \quad p \in \mathbb{Z}_k \\ \{\tau[A_{p+1}]\, x_0 \,|\, \mu\} &= \{x_{p+2} \,|\, \mu^{p+1}\}; \quad p \in \mathbb{Z}_k \end{aligned} \tag{9.31}$$

that, according to Eq. 9.27, and by taking into account the corresponding values i, j, can be written respectively from a common root as

$$\begin{aligned} \{\tau[-A_p]\, x_0 \,|\, \mu\} &= \{\mathfrak{R}^{p+2}\, \tau[-A_p]\, x_0 \,|\, \mu\} = \{\tau[\mu]_0^{p+2}\, x_0 - A_p \,|\, \mu^{p+2}\} = \\ &= \{x_0 + A_0 + \cdots + A_{p-1} + \cancel{A_p} + A_{p+1} \,|\, \mu^{p+2}\} = \{x_p + A_{p+1} \,|\, \mu^{p+2}\} \end{aligned}$$

and

$$\begin{aligned} \{\tau[A_{p+1}]\, x_0 \,|\, \mu\} &= \{\mathfrak{R}^{p}\, \tau[A_{p+1}]\, x_0 \,|\, \mu\} = \{\tau[\mu]_0^{p}\, x_0 + A_{p+1} \,|\, \mu^{p}\} = \\ &= \{x_0 + A_0 + \cdots + A_{p-1} + \cancel{A_p} + A_{p+1} \,|\, \mu^{p}\} = \{x_p + A_{p+1} \,|\, \mu^{p}\} \end{aligned}$$

Hence, in addition to Eq. 9.26, we get the following result.

▷ The following chord translations lead to the same co-cycle into a non-congruent cell,

$$\begin{aligned} \{\tau[-A_p]\, x_0 \,|\, \mu\} &= \{x_p + A_{p+1} \,|\, \mu^{p+2}\}; \quad p \in \mathbb{Z}_k \\ \{\tau[A_{p+1}]\, x_0 \,|\, \mu\} &= \{x_p + A_{p+1} \,|\, \mu^{p}\}; \quad p \in \mathbb{Z}_k \end{aligned} \tag{9.32}$$

While both translated chords of $\{x_0 \,|\, \mu\}$ in Eq. 9.26 are referred to the cell x_p, that is a neighbor cell of x_0, the translations of Eq. 9.32 are referred to a cell, that is beyond a neighbor cell by one mode interval, so to say, in a second ring of cells around the initial chord. Thus, the families of chords of Eq. 9.26 belong to one co-cycle in the cell x_p. However, both translated chords of Eq. 9.32 belong to one co-cycle in the cell $x_p + A_{p+1}$, not shared with the initial chord.

9.7 TRANSLATIONS TOWARD DIFFERENT CELLS

Chords belonging to different cells will mean that cannot be referred to the same cell. We study the other two possibilities of single translations.

First, let us see whether the families of translated chords

$$\begin{aligned}\tau[A_p]\,\{x_0\,|\,\mu\} &= \{x_{p+1}\,|\,\mu^p\}; \quad p\in\mathbb{Z}_k \\ \tau[A_q]\,\{x_0\,|\,\mu\} &= \{x_{q+1}\,|\,\mu^q\}; \quad q\in\mathbb{Z}_k\end{aligned} \tag{9.33}$$

assuming $p \neq q$, provide two chords that can be referred to the same root. By following the same procedure as the previous section, after rotating, the roots must match. Then,

$$A_{k-j}+A_{k-j+1}+\cdots+A_{k-1}+A_0+\cdots+A_{i-2}+A_{i-1}+A_p = A_q \mod n \tag{9.34}$$

Like in the previous case, this relationship has a geometric meaning. On the right-hand side there is one single positive interval. On the left-hand side, in addition to A_p, there are $i+j$ consecutive intervals, with $i+j \leq 2k-2$. Hence, altogether, to fulfill that relationship there must be one cycle plus one interval. If there is no cycle, then $p=q$, which is not the case.

Thus, the above relationship should be either

$$\underbrace{A_p+\cdots+A_{k-1}+A_0+\cdots+A_{p-1}}_{0 \bmod n}+A_p = A_q \mod n$$

or

$$A_p+\underbrace{A_{p+1}+\cdots+A_{k-1}+A_0+\cdots+A_p}_{0 \bmod n} = A_q \mod n$$

by yielding, in both cases, to $p=q$, against the prior assumption.

Therefore, translations of a chord by a single positive interval lead to different chord cells, irrespectively of the root value.

Second, let us see whether the families of translated chords

$$\begin{aligned}\tau[-A_p]\,\{x_0\,|\,\mu\} &= \{x_p\,|\,\mu^{p+1}\}; \quad p\in\mathbb{Z}_k \\ \tau[-A_q]\,\{x_0\,|\,\mu\} &= \{x_q\,|\,\mu^{q+1}\}; \quad q\in\mathbb{Z}_k\end{aligned} \tag{9.35}$$

provide two chords that can be referred to the same root, also by assuming $p \neq q$. By following a similar procedure, after rotating, the roots must match. Then,

$$A_{k-j}+A_{k-j+1}+\cdots+A_{k-1}+A_0+\cdots+A_{i-2}+A_{i-1}+A_q = A_p \mod n \tag{9.36}$$

We reach a similar result as the first case, leading to $p = q$, against the former assumption. Hence, translations of a chord by a single negative interval lead to different chord cells.

However, if the Tonnetz is degenerate and the interval vectors A_p and A_q have the same length, the cells $x_0 + A_p$ and $x_0 + A_q$, and the cells $x_0 - A_p$ and $x_0 - A_q$ have respective roots with a similar value.

9.8 HONEYCOMB OF TRICHORDS

The relationships obtained in the above sections allow us to interpret the honeycomb of trichords depicted in Figure 4.6 for the general case associated with a mode $\mu = [A, B, C]$, as well as those of Figure 4.8 for the particular case $\mu = [4, 3, 5]$.

Each chord is shared by 3 cells and can be referred to 3 different roots labeling the center of the congruent cells, corresponding to three roots of the three rotations of the directed chord.

Each cell has 6 principal facets: 3 front facets that are orthogonal to the vector intervals $+A, +B, +C$ in the cell a, and 3 back facets that are orthogonal to the vector intervals $-A, -B, -C$. For trichords, each cell is surrounded by $2^3 - 2 = 6$ cells, as many as facets. All of them are principal facets.

For example, the directed chord $a \,|\, [A][B, C] \sim \{a \,|\, \mu^0\}$ shares the facet orthogonal to the vector interval A in the cell a with the chord $a \,|\, [A][C, B] \sim \{a \,|\, -\mu^1\}$. This facet is common to the cells a and $a + A$, and the chords in this facet maintain the notes a and $a + A$.

From the other two rotations of $\{a \,|\, \mu^0\}$, i.e., $a + A \,|\, [B][C, A]$ and $a - C \,|\, [C][A, B]$, we determine the other two chords sharing a facet with $\{a \,|\, \mu^0\}$. For the former rotation, the directed chord $a + A \,|\, [B][A, C] \sim \{a + A \,|\, -\mu^2\}$ shares the facet orthogonal to B in $a + A$, that is, the facet common to the cells $a + A$ and $a + A + B = a - C$. For the latter rotation, the directed chord $a - C \,|\, [C][B, A] \sim \{a - C \,|\, -\mu^0\}$ shares the facet orthogonal to C in $a - C$, common to the cells $a - C$ and $a - C + C = a$ (according to Eq. 9.26, $\{a - C \,|\, -\mu^0\} = \{a \,|\, -\mu^2\}$).

As explained in Section 5.8, voice leading along facets in the tonal network is associated with paths that maintain the notes on either side of the interval to which the facet is orthogonal. This corresponds to paths along edges of the chord network. According to Section 7.5, in such a path it is possible to remain in the same cell or to divert to another cell through the edge connecting to a non-congruent cell. We apply these transformations, i.e., the combinations of transpositions and rotations of Eqs. 8.27 and 8.28 to the actual chord network.

Between the cells $a+A$ and $a+B$, the connected directed chords are $a\,|\,\mu^0$ and[5] $\mathfrak{D}_0(a\,|\,\mu^0) = a\,|\, -\mu^2$, which share the root a and the third note $a + A + B = a - C$ (the fifth in the case of $\mu = [4, 3, 5]$).

Between the cells $a + A + B$ and $a + A + C$, the connected directed chords are $a\,|\,\mu^0$ and $\mathfrak{D}_1(a\,|\,\mu^0) = a\,|\, -\mu^1$, which share the root a and the second note $a+A$ (the major third) with the original directed chord.

Between the cell $a = a + A + B + C$ and the non-congruent cell $a+A+B+A = a+A-C$, the connected chords are $a\,|\,\mu^0$ and $\mathfrak{D}_2(a\,|\,\mu^0) = a + A - C\,|\, -\mu^0$. They share the second note $a + A$ and the third note $a + A + B$ (the major third and the fifth) with the original chord.

In such a case, the movements along the edges of the tonal cell can be interpreted as being: $\mathfrak{D}_0$ a displacement toward the neighbor chord on the edge in the direction orthogonal to the mode interval C (which preserves the fifth interval); $\mathfrak{D}_1$ a displacement in the direction orthogonal to the mode interval A (which preserves the major third interval); and $\mathfrak{D}_2$ a displacement in the direction orthogonal to the mode interval B (which preserves the minor third interval) toward a non-congruent cell. These operators generalize Lewin's [26] basic operators P, R, and the leading tone exchange L of the neo-Riemannian triadic progressions. The elementary Hamiltonian circuit running all the chords in the cell results by the application alternatively of six consecutive operators $\mathfrak{D}_0$ and $\mathfrak{D}_1$.

On the other hand, the chords in one cell may be translated to neighbor cells and to the same cell. In the latter case, a translation is equivalent to a shift of the mode by maintaining the root. For $k = 3$, there are two mode classes and, therefore, two co-cycles that translations cannot mix, except in the case of a degenerate Tonnetz with a single mode class.

For the chords $\{a\,|\,\mu^i\}$ with $i = 0, 1, 2$, Eq. 9.4 with $j = 1$ provides the chords of one co-cycle, that are obtained as negative translations of other chords in the cell,

$$\begin{aligned} \{\tau_{-A}\, a\,|\,\mu^0\} &= \{a\,|\,\mu^1\} \\ \{\tau_{-B}\, a\,|\,\mu^1\} &= \{a\,|\,\mu^2\} \\ \{\tau_{-C}\, a\,|\,\mu^2\} &= \{a\,|\,\mu^0\} \end{aligned} \tag{9.37}$$

[5]In this case we obtain the mode $-\mu^2$, which belongs to the co-cycle of the mode $-\mu$, but in a general case with $k > 3$, we may reach other co-cycles which are not in such a relation with the mode μ, since there will be more than two mode classes.

They had been obtained in Eq. 4.10 and correspond to the chords marked with black circles in Figure 4.6. For $j = 2$, bearing in mind that $\tau_A\,\tau_B\,\tau_C = \tau_0$, Eq. 9.4 provides these chords from positive translations,

$$\begin{aligned}\{\tau_C\ a\,|\,\mu^0\} &= \{a\,|\,\mu^2\}\\ \{\tau_A\ a\,|\,\mu^1\} &= \{a\,|\,\mu^0\}\\ \{\tau_B\ a\,|\,\mu^2\} &= \{a\,|\,\mu^1\}\end{aligned} \tag{9.38}$$

Thus, each chord of one co-cycle in the cell a can be expressed from roots of the three congruent cells, which are the three rotations of the same chord,

$$\begin{aligned}\{a\,|\,\mu^0\} &= \{\tau_A\ a\,|\,\mu^1\} = \{\tau_{-C}\ a\,|\,\mu^2\}\\ \{a\,|\,\mu^1\} &= \{\tau_B\ a\,|\,\mu^2\} = \{\tau_{-A}\ a\,|\,\mu^0\}\\ \{a\,|\,\mu^2\} &= \{\tau_C\ a\,|\,\mu^0\} = \{\tau_{-B}\ a\,|\,\mu^1\}\end{aligned} \tag{9.39}$$

For the chords generated by the negative modes and its shifts, for $j = 1$ Eq. 9.10 provides the chords of the other co-cycle, obtained as negative translations of other chords in the cell,

$$\begin{aligned}\{\tau_{-C}\ a\,|-\mu^0\} &= \{a\,|-\mu^2\}\\ \{\tau_{-A}\ a\,|-\mu^1\} &= \{a\,|-\mu^0\}\\ \{\tau_{-B}\ a\,|-\mu^2\} &= \{a\,|-\mu^1\}\end{aligned} \tag{9.40}$$

and, for $j = 2$, it yields

$$\begin{aligned}\{\tau_A\ a\,|-\mu^0\} &= \{a\,|-\mu^1\}\\ \{\tau_B\ a\,|-\mu^1\} &= \{a\,|-\mu^2\}\\ \{\tau_C\ a\,|-\mu^2\} &= \{a\,|-\mu^0\}\end{aligned} \tag{9.41}$$

corresponding to the chords marked with white circles in Figure 4.6.

Thus, the chords in the cell a of the co-cycle generated by the inverse mode, expressed from roots of the three congruent cells are

$$\begin{aligned}\{a\,|-\mu^0\} &= \{\tau_{-A}\ a\,|-\mu^1\} = \{\tau_C\ a\,|-\mu^2\}\\ \{a\,|-\mu^1\} &= \{\tau_{-B}\ a\,|-\mu^2\} = \{\tau_A\ a\,|-\mu^0\}\\ \{a\,|-\mu^2\} &= \{\tau_{-C}\ a\,|-\mu^0\} = \{\tau_B\ a\,|-\mu^1\}\end{aligned} \tag{9.42}$$

We now center our attention to single translations of a chord toward the same co-cycle in the same or another cell. According to Eq. 9.26, for the chord $\{a\,|\,\mu^0\}$ we get[6] these following six chords, referred to the same cell a and the neighbor cells $a+A$ and $a-C$,

$$\begin{aligned}
p=0\,;\quad & \tau_{-A}\,\{a\,|\,\mu^0\}=\{a\,|\,\mu^1\}\,; & \tau_C\,\{a\,|\,\mu^0\}=\{a\,|\,\mu^2\}\\
p=1\,;\quad & \tau_{-B}\,\{a\,|\,\mu^0\}=\{a+A\,|\,\mu^2\}\,; & \tau_A\,\{a\,|\,\mu^0\}=\{a+A\,|\,\mu^0\}^{\dagger}\\
p=2\,;\quad & \tau_{-C}\,\{a\,|\,\mu^0\}=\{a-C\,|\,\mu^0\}^{\dagger}\,; & \tau_B\,\{a\,|\,\mu^0\}=\{a-C\,|\,\mu^1\}
\end{aligned} \tag{9.43}$$

The symbol † indicates translations that do not change the shift of the mode; therefore, they are chords referred to co-cells of $\{a\,|\,\mu^0\}$. The translations for $p=0$ were already included in Eq. 9.39 as chords belonging to the co-cycle of $\{a\,|\,\mu^0\}$ in the cell a.

Also, according to Eq. 9.32, for the same chord $\{a\,|\,\mu^0\}$ we get the following translations, which are the same chords as in Eq. 9.43, but now referred to non-neighbor cells of $\{a\,|\,\mu^0\}$, i.e., $a+B$, $a-B$, and $a-C+A$,

$$\begin{aligned}
p=0; \quad & \tau_{-A}\{a\,|\,\mu^0\}=\{a+B\,|\,\mu^2\}; & \tau_B\{a\,|\,\mu^0\}=\{a+B\,|\,\mu^{0\dagger}\}\\
p=1; \quad & \tau_{-B}\{a\,|\,\mu^0\}=\{a-B\,|\,\mu^0\}^{\dagger}; & \tau_C\{a\,|\,\mu^0\}=\{a-B\,|\,\mu^1\}\\
p=2; \quad & \tau_{-C}\{a\,|\,\mu^0\}=\{a-C+A\,|\,\mu^1\}; & \tau_A\{a\,|\,\mu^0\}=\{a-C+A\,|\,\mu^2\}
\end{aligned} \tag{9.44}$$

The relationships in Eqs. 9.43 and 9.44 may be summarized and completed with the remaining rotations of each directed chord as follows

	$\{a\,|\,\mu^0\}$	=	$\{a+A\,|\,\mu^1\}$	=	$\{a-C\,|\,\mu^2\}$
τ_A	$\{a+A\,|\,\mu^0\}$	=	$\{a+2A\,|\,\mu^1\}^{\ddagger}$	=	$\{a-C+A\,|\,\mu^2\}$
τ_B	$\{a+B\,|\,\mu^0\}$	=	$\{a-C\,|\,\mu^1\}$	=	$\{a-C+B\,|\,\mu^2\}^{\ddagger}$
τ_C	$\{a+C\,|\,\mu^0\}^{\ddagger}$	=	$\{a-B\,|\,\mu^1\}$	=	$\{a\,|\,\mu^2\}$
τ_{-A}	$\{a-A\,|\,\mu^0\}^{\ddagger}$	=	$\{a\,|\,\mu^1\}$	=	$\{a+B\,|\,\mu^2\}$
τ_{-B}	$\{a-B\,|\,\mu^0\}$	=	$\{a+A-B\,|\,\mu^1\}^{\ddagger}$	=	$\{a+A\,|\,\mu^2\}$
τ_{-C}	$\{a-C\,|\,\mu^0\}$	=	$\{a+A-C\,|\,\mu^1\}$	=	$\{a-2C\,|\,\mu^2\}^{\ddagger}$

Similar relationships can be obtained working from the inverse mode.

The chords in black are referred to the neighbor cells of the initial chord, where they form a full co-cycle (see Figure 4.6). The initial chord and both translations share one note. These cells are 0 steps away from the initial chord.

[6]In this case, $A_0=A$, $A_1=B$, $A_2=C$, $x_0=a$, $x_1=a+A$, and $x_2=a+A+B=a-C$.

The chords in gray are referred to the cells in the second ring of cells from the initial chord, containing only two chords of a co-cycle. The two translations share one note that does not belong to the initial chord, although each translation has a common note with the initial chord. These cells are 1 step away from the initial chord.

The chords in gray marked with the symbol ‡, are the only single translations of the initial chord in their respective cells. However, these cells, also in the second ring, are 2 steps away from the initial chord. Similar relationships can be obtained with the inverse mode.

As an example, the model has been explicitly applied to describe the chord network associated with the trichords of a non-degenerate Tonnetz and, in particular, with the mode that generates the major and minor triads. For k-chords in the 12-TET system, if $k = 3$ there are 14 non-degenerate Tonnetze, but for $k > 3$ there are only two, generated by the modes $[1, 2, 3, 6]$ and $[1, 2, 4, 5]$. Nevertheless, the current model can easily be extended to degenerate Tonnetze. In higher dimensions, degenerate Tonnetze of particular interest are those generated by a mode such as $[A, A, B, B, C, C]$, in order to study the hexachords in the 12-tone scale with the mode $[3, 3, 2, 2, 1, 1]$, whose permutations and submodes generate, in addition to the major and minor triads, the major and minor seventh chords $[4, 3, 4, 1]$, $[3, 4, 3, 2]$, the augmented and diminished chords $[4, 4, 4]$, $[3, 3, 6]$, their seventh chords $[4, 4, 3, 1]$, $[4, 4, 2, 2]$, $[3, 3, 3, 3]$, the dominant seventh chord $[4, 3, 3, 2]$, among others. Therefore, derived works from the current approach may include studying how the algebra of chords becomes restricted to less fine partitions of the octave, i.e., how it behaves with regard to submodes.

These and other similar geometrical and algebraic features can be used to generate sequences of chords, to determine voice leading criteria, or to set up composition algorithms. The work *Microcromia – 442 Colors* by Sara Cubarsi [16] provides such an example. As part of a set of pieces where a number of melodic and harmonic criteria lead the music flow, in this study three melodic voices visit, harmonically speaking, a series of chords generated by each one of the 442 mode classes of Table 2.3.

It is easily perceived that opening further the door of one visited subject, many more interesting problems are glimpsed that require far deeper mathematical tools.

This book proposes a way to understand the intrinsic structure of chords, and makes it available to everyone who wants to investigate this fascinating area of mathematics that is music.

Bibliography

[1] M. Andreatta, D.T. Vuza, and C. Agon. On some theoretical and computational aspects of Anatol Vieru's periodic sequences. *Soft Computing*, 8(9):588–596, 2004.

[2] C. Callender, I. Quinn, and D. Tymoczko. Generalized Voice-Leading Spaces. *Science*, 320:346, 2008.

[3] N. Carey and D. Clampitt. Aspects of Well-Formed Scales. *Music Theory Spectrum*, 11(2):187–206, 1989.

[4] N. Carey and D. Clampitt. Two Theorems Concerning Rational Approximations. *Journal of Mathematics and Music*, 6(1):61–66, 2012.

[5] N. Carey and D. Clampitt. Addendum to "Two Theorems Concerning Rational Approximations". *Journal of Mathematics and Music*, 2017.

[6] R. Chrisman. Identification and correlation of pitch-sets. *Journal of Music Theory*, 15(1/2):58–83, 1971.

[7] R. Chrisman. Describing structural aspects of pitch-sets using successive-interval arrays. *Journal of Music Theory*, 21(1):1–28, 1977.

[8] T.W. Christensen. *The Cambridge History of Western Music Theory*. The Cambridge History of Music. Cambridge University Press, Cambridge, 2002.

[9] J. Clough. Diatonic Trichords in Two Pieces from Kurtág's Kafka-Fragmente: A Neo-Riemannian Approach. *Studia Musicologica Academiae Scientiarum Hungaricae*, 43(3-4):333–344, 2002.

[10] R. Cohn. Maximally Smooth Cycles, Hexatonic Systems, and the Analysis of Late-Romantic Triadic Progressions. *Music Analysis*, 15(1):9–40, 1996.

[11] R. Cohn. Neo-Riemannian Operations, Parsimonious Trichords, and Their "Tonnetz" Representations. *Journal of Music Theory*, 41(1):1–66, 1997.

[12] R. Cubarsi. Harmonic distance in intervals and chords. *Journal of Mathematics and Music*, 13(1):85–106, 2019.

[13] R. Cubarsi. An alternative approach to generalized Pythagorean scales. Generation and properties derived in the frequency domain. *Journal of Mathematics and Music*, 14(3):266–291, 2020.

[14] R. Cubarsi. On the divisions of the octave in generalized Pythagorean scales and their bidimensional representation. *Journal of Mathematics and Music*, 18(1):1–19, 2023.

[15] R. Cubarsi. An algebra of chords for a non-degenerate Tonnetz. *Journal of Mathematics and Music*, 18(3):259–295, 2024.

[16] S. Cubarsi. Microcromia – 442 Colors. Available at `http://drive.google.com/file/d/0B8Q-t_r26gwLRGZrcUpHOFV6R1U/view?usp=sharing&resourcekey=0-h-Y-FcNuOFjsfRJidgyrxQ` and `https://soundcloud.com/sara-cubarsi/microcromia-442-colors/s-gjKLp`, 2015.

[17] J. De Loera, J. Rambau, and F. Santos. *Triangulations: Structures for Algorithms and Applications.* Algorithms and Computation in Mathematics. Springer Berlin Heidelberg, 2010.

[18] J. Douthett and P. Steinbach. Parsimonious Graphs: A Study in Parsimony, Contextual Transformation, and Modes of Limited Transposition. *Journal of Music Theory*, 42(2):241, 1998.

[19] D.S. Dummit and R.M. Foote. *Abstract Algebra.* Wiley, 2004.

[20] A. Forte. *The Structure of Atonal Music.* Yale [paperbacks]. Yale University Press, 1973.

[21] H. Fripertinger and P. Lackner. Tone rows and tropes. *Journal of Mathematics and Music*, 9(2):111–172, 2015.

[22] E. Gollin. Some Aspects of Three-Dimensional "Tonnetze". *Journal of Music Theory*, 42(2):195–206, 1998.

[23] H.v. Helmholtz. *On the Sensations of Tone as a Physiological Basis for the Theory of Music.* Longman's, Green, and Company, London, 1863. English translation by Ellis, A.J. 1875.

[24] F. Jedrzejewski. *Mathematical theory of music.* Collection Musique/sciences. Ircam-Centre Pompidou, Delatour, 2006.

[25] O. Kolman. Transfer principles for generalized interval systems. *Perspectives of New Music*, 42(1):150–190, 2004.

[26] D. Lewin. *Generalized Musical Intervals and Transformations.* Yale University Press, New Haven, CT, 1987.

[27] D. Lewin. Notes on the Opening of the F♯ Minor Fugue from WTCI. *Journal of Music Theory*, 42(2):235–239, 1998.

[28] B.J. McCartin. Geometric Proofs of the Common Tones Theorems. *International Mathematical Forum*, 10(6):289–299, 2015.

[29] B.J. McCartin. Geometric Proofs of the Complementary Chords Theorems. *International Mathematical Forum*, 11(1):27–39, 2016.

[30] O. Messiaen. Technique de Mon Langage Musical. 1. v. *Texte. Paris: A. Leduc*, 1944.

[31] J. Rahn. *Basic Atonal Theory.* Longman music series. Longman, 1980.

[32] Michiel Schuijer. *Analyzing Atonal Music: Pitch-Class Set Theory and its Contexts.* Eastman Studies in Music. Boydell & Brewer, 2008.

[33] J.A. Siegel and W. Siegel. Categorical Perception of Tonal Intervals: Musicians Can't Tell Sharp from Flat. *Perception & Psychophysics*, 21:399, 1977.

[34] D. Tymoczko. The Geometry of Musical Chords. *Science*, 313:72, 2006.

[35] D. Tymoczko. The generalized tonnetz. *Journal of Music Theory*, pages 1–52, 2012.

[36] A. Vieru. *Cartea modurilor (Book of modes).* Ed. Muzicala, Bucarest, 1980.

[37] M. Žabka. Algebra of harmony: Transformations of just consonances. In *Mathematics and Computation in Music: 6th International Conference, MCM 2017, Mexico City, Mexico, June 26-29, 2017, Proceedings 6*, pages 76–88. Springer, 2017.

Index

9781032982908